WORK IN THE SPIRIT
TOWARD A THEOLOGY OF WORK

Miroslav Volf

New York   Oxford
OXFORD UNIVERSITY PRESS
1991

Oxford University Press

Oxford   New York   Toronto
Delhi   Bombay   Calcutta   Madras   Karachi
Petaling Jaya   Singapore   Hong Kong   Tokyo
Nairobi   Dar es Salaam   Cape Town
Melbourne   Auckland

and associated companies in
Berlin   Ibadan

Copyright © 1991 by Miroslav Volf

Published by Oxford University Press, Inc.
200 Madison Avenue, New York, NY 10016

Oxford is a registered trademark of Oxford University Press

All rights reserved. No part of this publication may be reproduced,
stored in a retrieval system, or transmitted, in any form or by any means,
electronic, mechanical, photocopying, recording or otherwise,
without the prior permission of the publisher.

Library of Congress Cataloging-in-Publication Data
Volf, Miroslav.
    Work in the Spirit : toward a theology of work / by Miroslav Volf.
        p. cm.   Includes bibliographical references and index.
ISBN 0-19-506808-4
1. Work—Religious aspects—Christianity.   I. Title.
BT738.5.V64   1991   261.8′5—dc20   90-46022

9 8 7 6 5 4 3 2 1

Printed in the United States of America
on acid-free paper

To teta Milica

Preface

My interest in the question of human work dates back to 1980 when I started doctoral work at the Evangelisch-Theologische Fakultät of the University of Tübingen under supervision of Prof. Dr. Jürgen Moltmann. My task was to analyze and give a theological evaluation of Karl Marx's understanding of work (the dissertation was published in a shortened form with the title *Zukunft der Arbeit— Arbeit der Zukunft. Der Arbeitsbegriff bei Karl Marx und seine theologische Wertung*. München: Kaiser; Mainz: Grünewald, 1988). As I reflected on the problem of work over the years I became increasingly dissatisfied with the vocational understanding of work still dominant in Protestant circles. The vocational understanding of work was developed and refined in the context of fairly static feudalist and early capitalist societies on the basis of a static theological concept of *vocatio*.[1] Modern societies, however, are dynamic. A single, permanent, salaried, and full-time form of employment has given way to multiple and frequently changing jobs. Such a dynamic society requires a dynamic understanding of work.

It was clear to me that the dead hand of "vocation" needed to be lifted from the Christian idea of work. It is both inapplicable to modern societies and theologically inadequate. But what should replace it? The answer to this question I offer in this book occurred to me as I was preparing a position paper on the church as fellowship for the third quinquennium of the International Roman Catholic–Pentecostal Dialogue (1985–1989). As I was grappling theologically with the Pauline understanding of the church, I was reminded of the fact that Paul utilizes a very dynamic concept, *charisma*, for theological reflection on Christian activity (especially inside the Christian community). So I started entertaining the possibility of making the concept of *charisma*, developed theologically, the cornerstone of a theology of work. A thought that very soon acquired the self-evidence of a platitude emboldened me to start working on the project: since the whole life of a Christian is by definition a life in the Spirit, work cannot be an exception, whether that work is ecclesiastical or secular. *Work in the Spirit is one dimension of the Christian walk in the Spirit* (cf. Rom. 8:4; Gal. 5:16ff.).

The predilection for pneumatology was already present in my theological evaluation of Marx's concept of work. I suggested for the first time, however, a *pneumatological theology of work* based on the concept of *charisma* in my article "Arbeit und Charisma. Zu einer Theologie der Arbeit" (in *Zeitschrift für Evangelische Ethik* 31 [1987]: 411–33). Reactions to this article both by economists and theologians encouraged me to articulate with greater detail and precision the call I had issued for a shift from the vocational to a charismatic understanding of work.

Attempting a paradigm shift, even a miniature one as is here the case, is not only a fascinating but also a risky endeavor. Whatever one might think of the whole of Thomas Kuhn's theory of scientific revolutions and of its applicability to theology, he is certainly right that the initial formulation of a new paradigm always comes as a rough draft. Only in rare cases will the merciless scientific community refrain from throwing the draft into the scientific wastebasket and instead use it as a basis for further reflection. So I am well aware that my call for a shift from the vocational to a charismatic understanding of work is no more than a rough draft with an uncertain destiny.

To the critical eyes of scholars, drafts of synthetic visions often make for frustrating reading, even when these scholars have a sympathy for the vision. In order to formulate a synthetic theological vision of human work, one has to range not only over the diverse disciplines within theology (such as the doctrines of God and creation, anthropology, soteriology, pneumatology, ecclesiology, and eschatology), but also over the secular disciplines of philosophy, psychology, sociology, and economy. The breadth of my subject not only thrust me into fields of learning I cannot claim expertise in, but also forced me to leave many things assumed or merely suggested.

Because the shift I am suggesting is from the vocational understanding of work developed within the framework of the doctrine of creation to a pneumatological one developed within the framework of the doctrine of the last things, I expect that some readers will feel that my pneumatological and eschatological reflection is rather terse. They will, I suspect, want to find more about the Spirit's nature (person or energy), about the Spirit's relation to the other persons of the Trinity,[2] about the spheres of the Spirit's presence and activity (church or world), and about the nature of the Spirit's presence in each sphere. They might also ask theological questions of comparable weight about the underlying eschatology. For instance, the eschatological realism I am operating with unapologetically might appear strange or at least unfamiliar to readers distant either from the classical Protestant tradition or from the projects of political and liberation theologians.

Short of developing a full-scale pneumatology and eschatology, there is little I can do about this problem. And even that would probably not help. A full-scale pneumatology and eschatology might make these readers understand me better, but would probably not make them agree with me much more. For I suspect that at the level at which basic theological decisions are made, agreement takes place less through persuasion of the mind than through conversion of the heart (though I grant the interdependence of the two). I will follow what seems the only method appropriate to my task: I will develop theological concepts only to the extent that the need arises in the discussion of various aspects of the subject at hand. I can only hope that even those who inhabit different theological worlds than I will profit from my reflection, if for no other reason than by having somebody to disagree with.

Where I felt that the development of the underlying theological, philosophical, and economic views would sidetrack me from the main argument, I have tried in the notes to direct the reader elsewhere for justification and further elaboration of these views.

References cannot, of course, substitute for arguments. But when economizing space is necessary, they can serve the useful function of reducing mere repetitions of the persuasive arguments and positions of others. I encourage the reader to take the trouble at crucial points to consult the literature referred to in the notes. Also, my other publications (especially my book *Zukunft der Arbeit—Arbeit der Zukunft*) sometimes provide important supplemental arguments. Furthermore, the reader will do better to concentrate somewhat more on the whole picture than on its details. I do not mean to excuse in advance the inadequacies of the argument (which I do believe is a cogent one) but to remind the reader to evaluate the book in the light of its main purpose: the development of a new synthetic theological vision about human work.

The main point of the book—to call for a pneumatological theology of work based on the concept of *charisma*—is, of course, not dependent on all theological views expressed in the book, not even on all pneumatological and eschatological views. I take it that there is more than one theology of work possible that would be founded on the concept of *charisma*. The reader who disagrees with the way my theological tradition leads me to construct a theology of work might want to try using the same foundation to erect a building in which she would feel more at home. Though I would like to think that the building I constructed is built with solid materials and makes a comfortable home—indeed, that a different architectural solution would not make a better building—whoever takes the trouble to put up a different structure on the same foundation would not act against my purpose in writing this book.

Whitehead once said that the whole of Western philosophy is a series of footnotes to Plato. The Western economic theory and social philosophy of this century can be described as a series of footnotes to the great debate that Karl Marx started with Adam Smith. The debate, so it seems, abruptly ended in 1989. The "velvet" European Revolution decided it in favor of Adam Smith. Except in some conservative circles, however, Smith's victory lacks the flavor of a real triumph. People are increasingly aware that the tradition that Smith started does not have solutions to some major problems plaguing the world today (such as widespread abject poverty, dehumanization, and global ecological disaster). The Marxist brand of socialism being completely discredited, and capitalism being inadequte, concerned people worldwide are searching for a still-elusive third way. It is unlikely that this "third way" will in the near future be anything other than a "first way" strongly modified to take into account the contemporary social and ecological problems (a socio-ecological market economy). But it is not yet clear what these modifications should be, and there is not even agreement about the value system according to which these modifications should be made. My book is a small contribution to the complex search for the third way, a contribution written from one perspective (that of Christian theology) and about only one aspect of the issue (the problem of work).

While I was working on this book I was also involved in the international process of theological reflection about economic issues initiated at the Oxford Conference on Christian Faith and Economics in January, 1987. It culminated in the "Oxford Declaration on Christian Faith and Economics," which was released in January, 1990, at the second Oxford Conference on Christian Faith and Economics.[3] Since I was working as a drafter of that document in preparation for the conference, I had the advantage of having papers relating Christian faith and economics reach my desk from every continent. The stimulus and insights they provided are reflected in this book, even if I do not often quote this material. On the other hand, I incorporated into the draft of the document not only many ideas but also some key formulations of the manuscript of this book (which was at that time all but finished). A good deal of these formulations survived the criticism of one hundred theologians and economists, ethicists and development practitioners, church leaders and business managers from various parts of the world. I did not go back and change formulations in the original manuscript to avoid overlap. I was especially pleased to see the "Oxford Declaration" follow me in basing its theological reflection on work on the concept of *charisma* rather than vocation.

Since the book was researched and written not only in the United States but also on the European continent (Germany and Yugoslavia) I did not always have access to the existing English translations of the works I was using. As a result, some references—especially those to German philosophers such as Fichte, Hegel, Feuerbach, Marx, and Nietzsche—are to German editions. Tracing all the references in English translations would be not only laborious but impossible since I am presently in Tübingen on a research grant from the Alexander von Humboldt-Stiftung.

I taught the substance of the material presented in this book in two summer courses at Fuller Theological Seminary, Pasadena, California. I appreciated the inquisitive minds of a diverse body of students there—some of them psychologists, sociologists, and business people. Thanks go also to my teaching assistant at Fuller Theological Seminary, Wayne Herman, for collecting literature from various libraries in the Los Angeles area. Part of the manuscript was written in a lovely home in Santa Barbara overlooking the Pacific, which Drs. Schuyler and Ethel Aijian provided for my wife and me in the summer of 1988. The final draft of the book was finished during a sabbatical leave from the Evangelical Theological Faculty in Osijek, Yugoslavia. I put the final touches on the manuscript as a fellow of the Alexander von Humboldt-Stiftung. Special thanks go to my wife, Dr. Judith M. Gundry Volf, who took the trouble to read the whole manuscript, even though she was familiar with its contents through our stimulating conversations. Her penetrating questions sharpened my arguments, and if the book has any clarity, a good deal of it is due to her strong distaste for abstruse style. I would also like to thank the following people for help in reading the proofs and compiling the indexes: Marianne Bröckel, Ingeborg Schlecht, and especially Kirk Bottomly.

The book is dedicated to my nanny, Milica Branković-*teta Milica* ("Aunt Milica") as I still call her—who is now well into her eighties. "Angels do not toil," wrote Nathaniel Hawthorne, "but their good works grow out of them." Her angelic work, done with a love that I could not mistake for anything else, graced the first five years of my life. As much as anyone else I know, she worked in the Spirit, because in her eager longing for the coming new creation, she lived a life of "righteousness and peace and joy in the Holy Spirit" (Rom. 14:17).

Tübingen, West Germany   M.V.
September 1990

Acknowledgments

This book includes significantly reworked material from the following essays:
—"Arbeit und Charisma. Zu einer Theologie der Arbeit." In *Zeitschrift für Evangelische Ethik* 31 (1987): 411–33;
—"Human Work, Divine Spirit, and New Creation: Toward a Pneumatological Understanding of Work." In *Pneuma* 9 (1987): 173–93;
—"Market, Central Planning, and Participatory Economy: A Response to Robert Goudzwaard." In *Transformation* 4, nos. 3/4 (1987): 61–63;
—"On Loving with Hope: Eschatology and Social Responsibility," *Transformation* 7, no. 3 (1990): 28–31;
—"Materiality of Salvation: An Investigation in the Soteriologies of Liberation and Pentecostal Theologies." In *Journal of Ecumenical Studies* 26 (1989): 447–67.

I thank the editors of these publications for kindly granting permission to use material from these essays.

Contents

Introduction   3

Work in the Limelight   3
The Meaning of "Work"   7

   On Defining Work   7
   Work, Toil, and Employment   9
   Work as Instrumental Activity   10

Economic Framework   14

   Ethics and Economic Systems   14
   Democratic Planning   17
   Market and Planning   18

PART I   CONTEMPORARY WORLD OF WORK

1. The Problem of Work   25

Importance of Work   26
Transformation of Work   27

   From Agricultural to Information Societies   27
   Agricultural Production   28
   Manufacturing Production   30

Crisis of Work   35
   Aspects of the Crisis of Work   36
   Causes of the Crisis of Work   42

2. **Dominant Understandings of Work   46**

   Adam Smith's Understanding of Work   48

      Purpose of Work   49
      Division of Labor and Pursuit of Self-Interest   50
      Alienation   53

   Karl Marx's Understanding of Work   55

      Nature and Purpose of Work   56
      Alienation and Humanization of Work   58
      "The Great Civilizing Influence of Capital"   61

PART II TOWARD A PNEUMATOLOGICAL THEOLOGY OF WORK

3. **Toward a Theology of Work   69**

   A Theology of Work   71

      Work and Sanctification   71
      Work and God's Purpose with Creation   74

   On Crafting a Theology of Work   76
   Theology of Work and New Creation   79

      A Christian Theology of Work   79
      A Normative Theology of Work   81
      A Transformative Theology of Work   83
      A Comprehensive Theology of Work   84
      A Theology of Work for Industrial Societies   86

4. **Work, Spirit, and New Creation   88**

   Work and New Creation   89

      Eschatology and the Significance of Human Work   89
      Eschatological *Transformatio Mundi*   94
      Human Works in the Glorified World?   96
      Cooperatio Dei   98

Work and the Divine Spirit   102
   A Pneumatological Theology of Work?   102
   Work as Vocation   105
   Limits of the Vocational Understanding of Work   106
   A Theological Reflection on Charisms   111
   Work in the Spirit   113
   Spirit and Work in *Regnum Naturae*   117
A Christian Ideology of Work?   119
   God's Judgment of Human Work   120
   Work Against the Spirit   121

5. Work, Human Beings, and Nature   123

Spirit, Work, and Human Beings   124
   Work—A Central Aspect of Christian Life?   124
   Work and Human Nature   128

Spirit, Work, and Leisure   133
   Dream of a Leisured Society?   134
   Work and Worship   136

Spirit, Work, and Environment   141
   Work and Nature   141
   Spirit, Persons, and Nature   143
   Work as Cooperation with Nature   144

Spirit, Work, and Human Needs   148
   Expanding Needs   149
   Fundamental Needs   151

Excursus: Spirit, Work, and Unemployment   154

6. Alienation and Humanization of Work   157

Character of Alienation   158
   Job Satisfaction   158
   Nature of Workers and Character of Work   160

Reasons for Concern   161
   Alienating Work, Economic Progress, and Universal Emancipation   162

Alienating Work and Anticipation of New Creation   163
Christian Quietism?   166

Forms of Alienating Work   168

Autonomy and Development   170
Worker and Management   173
Worker and Technology   179
Work and the Common Good   186
Work as an End in Itself   195

Notes   203
Bibliography   229
Index of Scriptural References   247
Index of Authors   251

Work in the Spirit

Introduction

The rise and development of capitalism together with what has been termed—rightly or wrongly—the "Protestant work ethic" gave birth to the Western "world of total work."[1] Work has come to pervade and rule the lives of men and women, be it in the form of indefatigable or cruelly enforced industriousness, pure and simple, in the initial stages of the industrial nations' development; be it in combination with frantically pursued leisure at later stages. At first people worked incessantly because it was profitable, necessary, a matter of duty, or all three. Then they came to understand themselves as working beings whose highest destiny is to work and whose very being consists in the process of becoming something through work which they would not have been without it. It is no accident that both liberalism and its estranged younger sibling, socialism, centered their anthropologies and political theories around human work.[2]

WORK IN THE LIMELIGHT

Given the significance of work in modern societies, it is not surprising that people concerned about social health would turn their attention to the problem of work. R. Bellah's book *Habits of the*

Heart is a good example. After an analysis of the prevailing individualism in American culture (which shows striking similarities at this point with the cultures of most other economically developed countries), he asks, "Is it possible that we could become citizens again and together seek the common good in the postindustrial, postmodern age?"[3] In a brief answer to this question at the end of the book, a plea for the transformation of work figures prominently. Work "as a contribution to the good of all and not merely as a means to one's advancement," maintains Bellah, and work "that is intrinsically interesting and valuable *is one of the central requirements for a revitalized social ecology.*" Indeed, in a society that seeks the common good in a postindustrial age, such work should, in Bellah's opinion, be a "primary form of civic virtue." Such transformation of the meaning of work and our relation to work will not be achieved by "expert fine-tuning of economic institutions alone," but requires also "a deep cultural, social, and even psychological transformation."[4]

In recent years the question of work was propelled from its prolonged and undeserved backstage existence into the limelight of theological and ecclesiastical interest. The high unemployment rates in economically developed nations in the early eighties provided the occasion. The introduction of new labor-saving technologies created a strong impression that industrial societies were running out of work. For now, the impression has proven to be mistaken. Nevertheless, when people sense that work is a scarce resource in the world of "total work," they start to feel like they are being forced to breathe thin mountain air. Without work they are outside their life-sustaining environment. Because of the perilous consequences of the lack of work in societies dominated by work, church bodies all over the world in recent years have put the problem of work high on their agendas. Even if they were mistaken about the urgency and significance of the problem of unemployment, they were right in sensing the need for renewed theological reflection on the problem of work in rapidly changing technological societies.

In dealing with the problem of work, the Roman Catholic Church could draw on a long and honorable tradition of theological reflection about work and the social condition of workers. The tradition started with the publication of the encyclical *Rerum Nov-*

INTRODUCTION

arum (1891) written by the pope of the "social question," Leo XIII. It was a long-overdue official church answer to the plight of workers in the nineteenth century. Ninety years later, Pope John Paul II published the encyclical *Laborem Exercens*. In terms of content there is not much new in *Laborem Exercens* when compared with the older social encyclicals that followed after *Rerum Novarum*. But the *approach* to the question of work is radically different.[5] Previous encyclicals attempted primarily to alleviate a particular social evil, such as exploitation.[6] *Laborem Exercens* treats human work as a "perennial and fundamental" phenomenon and addresses it as "the *essential key* to the whole social question."[7] *Rerum Novarum* and subsequent social encyclicals were primarily social critiques. *Laborem Exercens* is primarily a philosophy and theology of work.

Laborem Exercens is one of the most remarkable ecclesiastical documents on the question of work ever written. It received widespread acceptance[8] (except among extremists on both ends of the ideological spectrum) and spurred renewed study of the question of work by individual theologians and various Roman Catholic national Bishops' Conferences. Of particular significance in the English-speaking world is the pastoral letter of the U.S. bishops, *Economic Justice for All: Catholic Social Teaching and the U.S. Economy* (1986), a document arising from the bishops' "preferential option for the poor," i.e., the commitment to be on the side of the poor and oppressed. The pastoral letter addresses primarily the problems of the U.S. economy, but its comments on the scourge of unemployment, the resulting pauperization, and a host of other economic topics are illuminating and highly relevant to other situations as well.

Though with less moral force than the Roman Catholic Church, Protestant Christian bodies in recent years have also addressed the problems of human work. The World Council of Churches (WCC) has, for instance, recently published various documents dealing with economic problems in general and work in particular. These documents have an important prehistory.[9] Between the two world wars, both before and during the period of the Great Depression, a good deal of theological reflection was going on in ecumenical circles about human work. It culminated in the *Report of the Section of Church, Community and State in Relation to the Eco-*

nomic Order, which was approved at the Oxford Conference on Church, Community and State (1937).[10] Although the report was written out of the experience of the Great Depression with its high unemployment rates, it did not sacrifice a broad theological perspective as it focused on combatting concrete problems. Because of its comprehensiveness, the report is a theological document that can still stimulate our thought and contribute to a solution for the problems of our day. The recent WCC documents on the question of work share the relevance of the report but lack a broad theological perspective. They deal primarily with the "tremendous question of unemployment" and analyze its causes: rapid technological change, the structure of economic life, or both; and they denounce unemployment and its consequences, such as loss of self-respect by the unemployed, and their growing poverty and powerlessness.[11]

After long years of neglecting social concerns, in recent years evangelical Christians have been vigorously involved in a "catching-up process" as they have come to much deeper appreciation that God's people have "*social* as well as evangelistic responsibilities in his world."[12] At the recent Oxford Conference on Christian Faith and Economics (1987), which was attended by a cross-section of evangelical theologians, economists, and business people, the problem of work was put on the evangelical agenda. They agreed to initiate a worldwide process of "further study on the following interrelated issues in the light of the scriptures: work; stewardship; creation of wealth; justice; freedom and democracy; leisure"[13]—a process that in the year 1990 culminated in a document entitled "Oxford Declaration on Christian Faith and Economics."[14]

It is reassuring to know that concerned Christian bodies are responding to the present crisis in work by striving to provide guidance for Christians who participate in economic life in various ways, as workers or managers, as employed, self-employed, unemployed, laid-off or retired persons. In order to be able to carry on this important task responsibly, Christian organizations will need to rely on careful theological reflection on the complex issues related to human work. In Protestant circles such theological reflection is, however, in short supply. The increased attention given to the subject of work has resulted in a number of popular theological treatments of the topic that often do not shy away from concrete proposals on matters of economic policy.[15] But as useful as such studies are for individual

Christians facing day-to-day decisions at their workplace, they cannot substitute for a more comprehensive and technical theological investigation of the problem of work in contemporary society, which Protestant theologians have yet to produce.

This book is an attempt to make a contribution to the alleviation of that dire deficiency in Protestant social ethics. Its purpose is to establish a broad theological framework for much-needed detailed theological and ethical reflection on the problem of work.

The purpose of a theology of work is to interpret, evaluate, and facilitate the transformation of human work.[16] It can fulfill this purpose only if it takes the contemporary world of work seriously. In Part I, I deal with the present reality of work by analyzing the character and the understanding of work in modern societies. After an initial methodological chapter, in Part II I then proceed with developing a pneumatological theology of work in an eschatological framework. But before I can devote myself to this main task I must indicate briefly the definition of work I will be operating with, and the economic framework my reflection on work presupposes.

THE MEANING OF "WORK"

On Defining Work

"If no one asks of me, I know; if I wish to explain to him who asks, I know not."[17] This was Augustine's predicament when trying to define "time." One feels much the same when attempting to define "work." Work is so close to us that nothing seems easier than to grasp what it is, yet our conceptual nets never quite manage to catch it.

The difficulty of defining work explains the odd "definition" of work in the opening pages of *Laborem Exercens*. There we read that work means "any activity by man, whether manual or intellectual, whatever its nature or circumstance."[18] Such a general statement about work cries out for a conceptual demarcation of work from other human activities. What follows, however, is an appeal to intuition: work means any human activity "that can and must be recognized as work, in the midst of all the many activities of which man is capable and to which he is predisposed by his very nature."[19] So work is finally whatever one thinks work is.

A part of the difficulty in defining work lies, certainly, in its ordinariness. Work is one of those things in our daily life "whose meaning is hidden in the mystery of their familiarity."[20] In addition, the character of work is presently undergoing a deep transformation due to technological innovations. The types of work that once dominated the world of work and were immediately associated with the term "work" are shrinking to insignificance, and new types of work are rising to prominence.

While an appeal to intuition is understandable when one deals with realities of everyday life that are in a state of flux, a theology of work cannot be satisfied with intuitions about its subject. Even within one culture, different people consider different activities to be work.[21] To avoid confusion, one must hence specify the understanding of work that underlies one's reflection. Such an act of specification is, of course, always partly arbitrary, since no one's specified use of a word is identical with its factual usage. But the arbitrariness is a problem only if the overlap between the term as it is defined and as it is intuitively understood becomes too small. In that case, the definition of a term *creates* confusion rather than eliminating it.

The avoidance of a denotative confusion is one reason for the need to reflect on the meaning of "work." Words, however, are not merely tools for communication, but are also often instruments of oppression. It makes a difference, for instance, whether we acknowledge doing household chores as work or not. In most economically developed societies, when we say that someone is "working" we not merely indicate that a person is involved in a particular kind of activity but also implicitly ascribe value to that person. To work is good; not to work is bad. A person who does not work is less valued in a society. Thus we hear deprecative remarks about women (or even descriptions by women of themselves) who are "only housewives." Furthermore, work gives access to monetary power and hence to independence, and is a door to active participation in the home and the society at large. Because their activity is not valued and remunerated at work, housewives are often barred from participating in decision-making that influences their lives. Finally, not to call doing household chores "work" helps hide exploitation of many women in modern societies. For they not only get lower pay on the average than men do for the same work, but also put in an additional fifteen to twenty-seven hours a week of housework—significantly more than men do—that is not valued as work at all. A responsible theology of work must operate with a definition of work that does not lend itself to such oppressive misuse.

Work, Toil, and Employment

In colloquial use the word "work" generally means either toil and drudgery or gainful employment.[22] Both meanings are largely correct descriptions of work, but are obviously inappropriate either as formal definitions of work or as components of a formal definition of work. Both share the fallacy of considering culturally and historically conditioned dominant characteristics of work as inherent in work itself. What we need to do in defining work is not to indicate what characteristics work can have or even for the most part does have, but to point out what characteristics an activity *must* have in order to be considered work, and what features differentiate work from other activities.

That work means drudgery is deeply rooted in the etymology of the word "work" in Indo-European languages. In Croatian, my own mother tongue, the words for "work" (*rad*), for "slave" (*rob*), and for "prisoner" (*robijaš*) are, for instance, closely related etymologically.[23] The English word "labor" may have the same etymological root as "labare," which means "to stumble under a burden." The understanding of work as drudgery also corresponds to the experience of the great majority of working people throughout history. And although people today are increasingly coming to believe that work should be fun, there is no doubt that most workers still work in the sweat of their brows. But drudgery is certainly not a necessary characteristic of work. For then we would have to make the absurd claim "that what the most capable worker does is no longer work,"[24] since the most capable workers as a rule enjoy their work the most.

The prevalent way we use the word "work" today is to designate gainful employment. After the industrial revolution, gainful employment became the main way of providing for sustenance for the majority of the population and hence came to occupy a central place in the lives of individuals and their families.[25] Since gainful employment was the most dominant form of work it became identified in people's minds with work. But the two are not identical, not even in modern economies, where there is a close tie between working and earning. For neither is all employment "work" (a person can be gainfully employed without actually working) nor is all work "employment" (a person can work without receiving any compensation). Moreover, the understanding of work as gainful employment tends to be alienating, because by definition it gives work the primary meaning of earning, not working. As I will show later, the more a person works for the sake of working, the less alienated she can be said to be.[26]

As I have indicated earlier, not to consider doing household chores "work" (which is a necessary consequence of the reduction of the concept of work to gainful employment) is oppressive for women. It is moreover arbitrary, since the same activities, if done in someone else's household, would count as work! The reduction of the meaning of work to gainful employment is also inadequate in the context of the Two-Thirds World. Its limitations become obvious if we apply it in the case of a country like Ghana. In the mid-1980s, the population of the country was about twelve and a half million. About 500,000 people were engaged in formal, recorded wage employment. Up to a million Ghanaians were self- or family-employed in the informal sector (small-scale food preparation, tailoring and shoemaking, utensil- and furniture-crafting, etc.). But the majority of the active population (some four million) were self-employed in agriculture, primarily for household survival.[27] If we stick to the understanding of work as gainful employment, we are forced to make the absurd claim that the majority of the economically active population of Ghana does not work.

Work as Instrumental Activity

In order to do justice to the changing reality of work in the modern world and avoid the oppressive implications of inherited concepts of work, I have proposed the following comprehensive definition of work: *Work is honest, purposeful, and methodologically specified social activity whose primary goal is the creation of products or states of affairs that can satisfy the needs of working individuals or their co-creatures, or (if primarily an end in itself) activity that is necessary in order for acting individuals to satisfy their needs apart from the need for the activity itself.* I have elsewhere elaborated on this definition of work and do not intend to repeat that discussion here.[28] But I do want to make three comments about this definition: first, a short one about its nature; second, a long one about the way it distinguishes work from other human activities; and a third one about its implications for the scope of this book.

First, this definition is a purely *formal* definition of work. It suggests what characteristics an activity that should be considered as work must have. It does not say anything normative about the character of work; for instance, what kinds of work are compatible with human dignity and what kinds are not.

Second, I have followed the standard procedure and defined work in such a way that leisure is opposite to work[29] (without, however, wanting to imply that the two poles of work and leisure encompass the totality of human life activity,[30] as I am not sure that such activities as eating fit into either category). Often work and leisure are similarly understood in a polar way, but the polarity is taken to be one of coercion (work) vs. freedom (leisure).

Various suggestions have been made about how to understand the coercive nature of work. Some claim that work necessarily includes social subordination. "The essence of work," says Heilbroner, is that objectively defined tasks "are carried out in a condition of subordination imposed by the right of some members of society to refuse access to vital resources to others."[31] This definition is not very persuasive, since it arbitrarily excludes a host of activities from the concept of work (homemaking by a single parent, for instance, could never be work according to this definition). According to a commoner view, the coercion that is integral to work does not necessarily come from the social context, but is integral to the work process. "Even the worker who is free in the social sense . . . feels this compulsion, were it only because, while he is at work, his activities are dominated and determined by the aim of his work. . . . Work inevitably signifies subordination of the worker to remoter aims."[32] But such an understanding of coercion erases, in effect, the distinction between work and leisure activity, for in every engaging hobby a person subordinates herself to remoter goals. Work and freedom are not mutually exclusive. Work is not necessarily a sacrifice of freedom, but can be and truly is at its best an exercise of freedom.[33] Of course, work can never be a completely free activity. No significant human activity is completely free; not even thinking. Even when a person is working under no outside compulsion, his or her work is always determined in many ways: through the purpose that needs to be accomplished, through the obstacles that lie in the way of its realization, and through the more-or-less set methods used to overcome those obstacles.[34] But if work is done from an inner need, at one's own pace and using one's own skill, a person can be as free in work as in any other significant human activity.

The essential characteristic of work that distinguishes it from leisure is not outer or inner coercion, but *instrumentality*. An activity called "work" is a means to an end that lies outside that activity itself. Yet it would be too simple to define work as "an instrumental activity carried out by human beings, the object of which is to preserve and maintain life."[35] For not every instrumental activity is work and all work is not experienced primarily as an instrumental activity. A hobby, which by definition is a non-work activity, may be an instrumental activity. It is rather useful to have a person in the home whose hobby is doing minor repairs. But no sooner does the primary goal of that person become, not enjoying the activity of repairing itself, but solving the problems caused by things' getting broken, than that person's hobby turns into (more or less enjoyable) work. On the other hand, an activity can be experienced as noninstrumental activity (i.e., it may be, subjectively speaking, an end in itself) and still be work, if it is objectively necessary to satisfy the needs of the working individual (apart from the need for this activity) or of the people for whom she is responsible. It is imaginable, for instance, that a gardener can work more for the sheer pleasure of gardening than for the remuneration she might get from gardening. But if her work is not to be transformed into a useful hobby, then gardening has to be necessary for her to satisfy her needs. If the work is not objectively necessary to satisfy needs it can still be partly an end in itself and still be considered work, provided that it is not done primarily for its own sake, but to satisfy needs. In short, what distinguishes pleasant work from a useful hobby is that work must be either necessary to satisfy needs other than the worker's need for the activity itself or be not primarily done for its own sake.

My definition of work as an instrumental activity serving the satisfaction of needs should not mislead anyone into thinking that the definition is individualistic: the test of whether an activity is work is whether it satisfies a worker's needs. The needs I am talking about, however, are not only the needs of working individuals themselves but also *the needs of "their co-creatures."* The somewhat cumbersome term "co-creatures" is meant to include the whole human race as well as living nonhuman creatures. In this way the need for doing socially and ecologically responsible work is reflected in my definition of work.

As I have defined work, the distinction between work and hobby need not be made on the basis of the objective characteristics of the activity performed; it can be made on the basis of the subjective attitude of the person performing it. The same activity can be both work and leisurely activity, depending on whether a person does it primarily as a means or as an end in itself. This subjective component in the distinction between work and leisurely activities means that the line separating work from leisurely activities is blurry and that in some situations it will be impossible to make a distinction between the two. This impossibility does not stem from conceptual imprecision, however, but rather reflects the actual human experience in borderline cases.

Third, the broad definition of work as an instrumental activity for satisfying the needs of human beings (and other creatures!) implicitly determines the scope of this discussion. It is not a book on drudgery or on gainful employment. I intend to reflect theologically on most of what, in Christian tradition, has been termed *vita activa* (leisure activity excluded).

On the one hand, this intention seems natural: insofar as particular types of human activities (such as economic, social, political, cultural, and ecclesiastical activities) are specific instances of general instrumental activity, they also need to be studied in the context of theological reflection on instrumental activity in general. On the other hand, the attempt to reflect theologically on the whole range of instrumental activities clashes with a marked tendency in modern societies for various spheres of human life to function according to their own logic. Having differing operational logics, these spheres tend to develop their own specific moralities: one morality for economic life, for instance; a different morality for private life.

Christian faith, however, is incompatible with a plurality of moralities in one life of commitment to God, for by definition, God is a God of the whole of reality. For a Christian the answer to the question of how to function responsibly in each particular sphere of life depends on the answer to the question of how one should responsibly live one's life as a whole. My life project cannot be simply a result of my ability to integrate various independent spheres I function in. The various functions I perform must conform to the notion of the "good life" lived before God. Therefore, reflection on particular instrumental activities presupposes reflection on instrumental activity in general. I am, of course, not denying the need for specific reflection on different types of human work, such as industrial, agricultural, medical, political, or artistic work. I am only saying that insofar as each of these types of work is *work*, general theological reflection on the meaning, nature, and purpose of work applies to all of these equally.

ECONOMIC FRAMEWORK

The comparative study of economic systems and their accompanying ideologies is a complex task prudent economists and social scientists shy away from. I would have followed suit, were it not for the need to specify in what economic framework the following theological reflection on work should be placed. The positive form of my discourse notwithstanding, my intention is here more negative: I will only try to indicate what economic system my reflection of work is incompatible with; I will not try to suggest what an optimal economic system should look like.

Ethics and Economic Systems

The claim that some economic systems might be incompatible with theological perspectives suggests that economic systems should follow the dictates of theological reflection rather than that theological reflection should follow the functional logic of a given economic system.[36] This suggestion betrays my belief that the evaluation of economic systems is a task proper to theological ethics. Economists today tend to think differently. For some time now, an increasing tendency toward the autonomisation and desocialisation of economic activity and economic science has been visible.[37] Economic activity is guided by the principle of economic efficiency, and part of the job of an economist seems to be to make sure that no "subjective" ethical norms and value judgments interfere with the process. But economic systems are only apparently value-free. There are always certain values that not only guide individual economic choices but are also embodied in economic institutions.[38] Every socioeconomic system has built-in implicit normative presuppositions. Hence the possibility and the need for ethical reflection on economic systems.

True, very often ethical demands fall on deaf ears: the players of the economic game follow the internal rules of the economic system rather than the external demands of ethical norms.[39] For ethical norms are often reinforced only by individual conscience. Economic rules, on the other hand, are reinforced by the tangible consequences of pecuniary gain or loss, which can often mean the life or death of the economic players. A person certainly *can* hold to ethical norms,[40] but often only so long as the dynamic of the system does not crush him. Faced with the choice between obedience to conscience and survival, he is likely to opt for survival.

If the distinction between the knowledge of good and the realization of good holds true (which I believe it does), then the ineffectiveness of ethical norms in a particular sphere of human activity is not a sufficient reason to be unconcerned about them. It is rather an incentive to search for better ways of bringing moral discourse to bear on the functioning of economic systems, a search that will have to include a careful investigation of when one can trust the ethical decisions of individuals and when one should strive to introduce legal restrictions.[41]

I believe that economic systems should be judged primarily by three normative principles: freedom of individuals, satisfaction of the basic needs of all people, and protection of nature from irreparable damage. All three of these principles can be derived from the notion of "new creation," which in the present book functions as the main ethical norm. First, the concept of new creation implies *guarding the individual's dignity*. Each person is created in the image of God and is called to a personal relation with Christ as His brother or sister. In economic life the individual is thus not to be treated as a thing but as a *free and responsible agent*. Second, the concept of new creation has implications for *community*. It implies *practicing solidarity*. Every person is called to be an heir with Christ in the community of God's people. In economic life this should mobilize us to *work for the fulfillment of everyone's basic needs*. In particular, it implies a preferential option for the poor. Third, the concept of new creation has implications for the *natural environment*. It implies the *responsibility of preserving the integrity of nature*. Nature is not a mere thing. Rather, the Bible teaches us that it suffers under corruptibility and will participate in the freedom of the children of God (Rom. 8:78ff). In economic life this implies that *protection of nature from irreparable damage* must accompany any work on or in nature.

I recognize that translating normative principles into operational policies is a complex and difficult task in which one has to take into account not only normative guidelines but also the complex historically and culturally conditioned reality in which these guidelines take on concrete form; but I would still venture to say that all three of these principles require *both a market and planning* as essential elements of a responsible economic system.

The mention of the necessity of planning will come as a surprise to some people. Planning is not very popular today. Whereas a few decades ago socialism, with its emphasis on government intervention and planning, was the order of the day, today for numerous reasons capitalism has become the spirit of the age: markets are increasingly left to pursue their course, and entrepreneurship elicits praise—even in socialist countries. All present economic reforms in the socialist countries—whether in China, the Soviet Union, or Yugoslavia—involve either introducing elements of a market economy or allowing more freedom for the operation of the market. In the book *Perestroika* Gorbachev, for instance, summarizes his proposed economic changes by indicating his intention to increasingly combine the advantages of the planned economy with the stimulating factors of the socialist market.[42] Recently he has moved even more radically toward a free market economy. The present worldwide fascination with the free market notwithstanding, I believe that today's economic systems can function in a healthy and responsible way only if the market is checked by planning. But before I give a schematic (but not exhaustive) list of reasons for the necessity of both market and planning, I need to explain briefly what I mean by "economic planning."

Democratic Planning

Whether economic planning results in occasional interventions in the operation of the market or (preferably) in the setting up of a stable legal framework for economic activity, economic planning can be an integral part of a responsible economic system only if it is characterized by the following features:

First, planning should not be seen as a substitute for the market but as a supplement to it.[43] At the same time, it should not be limited to balancing out the autonomous economic subsystems, but should be understood as exogenous goal-setting based on normative guidelines and reflection on how to implement those goals most efficiently for the purpose of compensating for the anthropologically, socially, and ecologically detrimental consequences of the market. Although compatibility with the market might be a desirable feature of planning, it is not a necessary one. But it is necessary that both market and planning be compatible with normative guidelines.

Second, since direct or indirect intervention is simply a more or less useful instrument for correcting the market in the light of normative principles, planning should be exercised only to the extent demanded by these normative principles. There is no virtue in government intervention as such. In fact, there is a danger in it, because all social planning necessarily involves limiting individual freedom. The bitter experiences in the East and the West should have dispelled all illusions that a government apparatus is a neutral instrument that can be used to achieve certain socially desirable goals.[44] Government intervention is therefore legitimate only when other important values are threatened. The principle of subsidiarity, which maintains that a higher level should not do for a lower level what the lower level can do for itself, should be applied to government planning.

Third, direct or indirect steering of the market mechanism must be legitimized and controlled by democratic processes. The emphasis on the democratic nature of planning has special importance, not only because of the predominantly nondemocratic nature of all centrally planned economies, but also because new developments in communication technology make possible the government's tutelage of the people, even their total surveillance.

Finally, false expectations with which planning was overloaded since the French Revolution must be shed. That revolution gave birth to a rather dangerous illusion that institutional changes, if carried out properly, can solve all the problems of humankind.[45] Social and economic planning is not a means of salvation in this world. It is an "open, explicit, rational goal-setting process"[46] which takes into account the means of achieving the goals set and the potential negative side-effects under the direction of guidelines understood as normative rather than absolute.

If one understands planning in this way, then I propose with Alperowitz that not to plan "is absurd on the face of it. In our daily lives, in our families, in our communities, in our corporations, in our local and national institutions it is obvious that we must look to the future and plan ahead. Only in our economy do we say this is a mistake. All the while we quietly continue a trend of planning which is done badly and deceptively in significant part because of our lack of candor. It is time to recognize the reality—and common sense. The issue before us is not whether to plan but *how* to plan effectively, by which groups, and toward which goals."[47]

Market and Planning

All three normative principles mentioned above—freedom of individuals, satisfaction of the basic needs of all people, and protection of nature from irreparable damage—require *both* planning and market as elements of an economic system.[48]

First, with respect to *individual freedom*, elements of the *market* seem necessary to guarantee freedom of production and consumption (though one should have no illusions about producers' manipulative power over consumers). A centrally planned economy that allows only a marginal role for the market results in alienation in work and tyranny over human needs, because the choices of the economic actors are severely limited by bureaucratic decisions the economic actors have virtually no control over.

Elements of democratic *planning*—i.e., (preferably) indirect steering of the market mechanism—are necessary in order to give individuals the opportunity to influence the overall conditions and content of their economic activity as well as to be able to make significant decisions at their place of work. Traditional market economies put all the power in the hands of the owners and thus make the majority of the population unable to exert significant influence on one dominant aspect of their lives, the economic aspect.

The possibility of the active participation of workers at their workplace seems to diminish with the increase in the company's size. Small is not only beautiful, but also more conducive to participation. To keep companies small or to make them small, the element of democratic planning may be necessary. Furthermore, as neoliberals have come to recognize, a competitive market is not a natural state of affairs but needs to be guarded by state intervention.

Second, with respect to *meeting the basic needs* of every individual, the *market* is necessary because it secures greater economic efficiency (not only through competition but also through the greater potential for technological innovation based partly on competition). There are definite "social benefits" of the market, which one must not forget while enumerating its "social costs." As most socialist countries in the world today are recognizing on the basis of their own prolonged bitter experience, inflexible bureaucracy in centrally planned economies significantly reduces economic efficiency. Even for the best and most benign bureaucracy—assuming there is such a thing—it is technically impossible to acquire all the information necessary to make correct economic decisions. Moreover, even if the bureaucracy does make correct decisions, it has no adequate means to ensure their being executed responsibly and efficiently. Part of the reason is the difficulty socialist planned economies have in properly placing blame when things go wrong.[49]

Some form of central *planning* is necessary, however, in order to meet the basic needs of all people, not only because planning seems to be necessary to provide stability to the economy but also because the market alone has no mechanism for the redistribution necessary in modern societies if the needs of disadvantaged people are to be met.

Planning on a *national level* is necessary in order to (1) guarantee quality in public services, (2) orient production more to the needs of the relatively poor majority than to the rich minority, (3) solve the problems of structural unemployment, and (4) guarantee a decent life to the growing population of the elderly unemployed. Planning on an *international level* is necessary both to solve economic problems created by the population explosion and to prevent the flow of wealth from poor to rich nations.

Third, with respect to the protection of the *integrity of nature* from irreparable damage, the *market* seems to be more efficient in the use of existing limited resources. These are too precious to be wasted by a bureaucracy that is inefficient and workers who are irresponsible because they are uninterested.

Planning is, on the other hand, necessary to prevent the indiscriminate exploitation of nature motivated by individual or corporate desire for profit. In the long run, the market seems to be blind to the future, and yet through the market, irrevocable decisions are being made that affect the future of the whole human race.

Conservative Christians in the Western world prefer the combination *"evangelism and market"* to the combination "planning and market" that I am suggesting. This reflects their dislike for structural change and their enthusiasm for individual change.[50] I have no quarrel with what they affirm (the need for individual change), but I do with what they deny (the need for structural change and for planning). With them I want to stress the importance of individual change not only for restoring a person's relation to God but also for enabling the proper operation of the economy. Restructuring in economic life must be accompanied by transformation in personal life. Only then can an economy be healthy. *Both market and democratic planning* (as does democracy in general) presuppose morally responsible individuals. One of the many ways to make this point is to refer to the devastating influence of corruption on economic life.[51]

Some economists might object that in suggesting combining market with planning I am trying to have my cake and eat it, too: I want the advantages of both market and planned economies without their disadvantages. But very few economists today would contrast market and planning so sharply as L. von Mises did when he claimed that "there is simply no other choice than this: either to abstain from interference in the free play of the market, or to delegate the entire management of production and distribution to the government. Either capitalism or socialism: there exists no middle way."⁵² In fact, in all existing economies in the industrial world today, market and planning coexist inseparably in the same economic system. So the question is not whether we should combine market and planning, but how we should do it so as to be sensitive to the principles of the freedom of every individual, the preferential option for the poor, and the need for the protection of nature from irreparable damage.

PART I

Contemporary World of Work

CHAPTER 1

The Problem of Work

A theology of work is a critical theological reflection on the reality of human work. Since its purpose is both to evaluate the world of work and to participate (in its own way) in reshaping the world of work in the light of the promised new creation,[1] the first step in developing a theology of work must be to study the present reality of human work. A theologian needs to learn how people go about doing their work and how they have come to interpret it. Both the activity of work and its interpretation belong inseparably to the reality of human work. In the following chapter (Chapter 2) I will discuss the two dominant ways of understanding work today. Chapter 1 deals with the activity of working itself: its anthropological and societal importance, its rapid transformation through history, and its deep present-day crisis.

The purpose of analyzing the contemporary world of work is both to highlight the significant features of work that a theology of work has to take into account, and to specify the unresolved problems it needs to find solutions for. The relation between problems and solutions in this discussion is not that of questions and concrete answers. For instance, I will not take each aspect of the crisis of work and indicate how they are resolved by a theology of work I am proposing. Neither will I take issue with each claim of the dominant understandings of work by showing how the theology of work I am proposing is more persuasive. Instead, while focusing on developing a theology of work in Part II, I will keep the analysis of the contemporary world of work from Part I in my peripheral vision. In other words, I will be taking the reality of work into account without *directly* responding to each of its challenges.

IMPORTANCE OF WORK

We take work for granted almost as much as we do the air we breathe. Except when we are laid off, retire, or become incapacitated for work, we rarely pause to think either of the significance of our own work for ourselves or of the combined work of all human beings throughout history for the life of humanity. If the whole world decided to take an extended common holiday—say, for a few months—and did absolutely no work, it is not hard to predict what would happen: the world population would perish before the holiday terminated. The apostolic injunction, "If anyone will not work, let him not eat" (2 Thess. 3:10—until recently part of the Soviet constitution!) rests on the simple underlying principle, "If nobody works, nobody *will* eat." Human beings do not live by bread alone; that is what Jesus tells us. Nobody needs to remind us, however, that we do not live without bread either, and that we eat it only by the sweat of our (or somebody else's) brow.

The significance of human work, however, goes far beyond providing human beings with the necessary means of sustenance. We not only live from what we do, but to a large extent, we also *are* what we do. Although there is an important sense in which this statement is not true,[2] one can hardly deny that we cannot understand ourselves anthropologically (i.e., who we are as human beings) and sociologically (how our societies are structured and how they function) without taking into account the ways in which we go about doing our daily work. Claims like this are generally associated with the name of Karl Marx. As is well known, this indefatigable (though by no means unbiased, to say the least) student of the world of work maintained that who people are coincides with the conditions and modes of their material production.[3] Marx may have worked out and popularized this anthropological and sociological view, but he did not originate it. It was Adam Smith who fathered the double thesis that who people are is to a large extent determined by what they do and that how societies are organized is fundamentally influenced by their basic mode of production.[4]

Work is indispensable for the survival and the well-being of both individual human beings and the societies they live in, and it conditions their individual and social identity. As such, it is the basis of individual human life and of all human history. Christians, of course, rightly maintain that the whole course of human history is a result of God's preserving and guiding activity. At the same time we must acknowledge that we can read—indeed, as the Yugoslav poet Aleksa Šantić puts it, that *God*, full of joy, can read—the whole of human history in the "calloused hands" of diligent workers.[5]

TRANSFORMATION OF WORK

Even though work is a fundamental condition of human history, the nature of human work does not remain unaltered in different historical periods. During most of human history, of course, the wisdom of a trade was passed virtually unchanged from generation to generation. Any transformations of the conditions and character of work were spread over the centuries. But after industrialization (and especially with the discovery of computer technology in the second half of the twentieth century), the world of work has been permanently revolutionized by increasingly rapid technological development.

From Agricultural to Information Societies

We can roughly divide the history of human work in economically developed industrial nations up to the present into three consecutive eras: the agricultural, the industrial, and the information (and service) eras.[6] Throughout history people have, of course, produced food, manufactured goods, and processed information, all at the same time. They will continue to do so in the future, for these are the three basic fields in which human work must be done. Nevertheless, the division into periods is helpful since a particular type of work dominated in a given historical period. In ancient times (but today also in many parts of the world), very few workers could be spared for nonfood tasks. In economically highly developed countries, like the United States, on the other hand, 3 percent of the population produces more than 100 percent of the required agricultural goods.[7] In ancient times, only a small, privileged elite could devote themselves to "processing information." Today, a not-so-privileged-majority of the population (well above 60 percent in the United States, with the percentage rapidly on the increase) holds various information jobs.[8]

The statistics that show a drastic decrease of agricultural and increase of information tasks signal a radical transformation in the world of human work. At present this change is taking place, for the most part, in economically developed nations. Many developing nations are still predominantly agricultural or manufacturing societies. But the future of work in these societies, too, will be marked by a (more or less rapid) movement away from the agricultural to the information tasks.

The transition from the agricultural to the information society has been made possible only through the radical changes taking place in each of the basic fields of human work (agriculture, manufacture, and processing information). I will illustrate this point in the following sections by making a comparison between food production and manufacturing in ancient (and many contemporary developing) societies on the one hand; and in modern, economically developed societies, on the other hand.

Agricultural Production

I will depict the radical changes in agricultural production partly on the basis of my firsthand observations in Yugoslavia. The farm life I experienced as a child some twenty-five years ago corresponds fairly accurately to farm life, not only in many contemporary developing countries, but also in the ancient world.[9] Present-day farming on large agricultural complexes in Yugoslavia also exemplifies in many respects modern agricultural methods used worldwide.

Small-Scale Farming

In Sirač—the farming community I frequently visited as a child—farming was done on small family farms surrounding the village.

Women (with the aid of children) took care of domestic duties, including baking bread, feeding cattle, and tilling the vegetable and flower gardens. Men worked in the fields (with women helping them in high season). To reach the fields they would either walk or ride on a cart with wooden wheels pulled by a pair of horses, a trip that often took the better part of an hour. The land was tilled by horses pulling a small iron plow as a plowman pressed it into the hard soil. Sowing was done by hand, the seed carried in an apron and scattered on the ground. Animal dung was collected and carried on a cart to be distributed through the fields. The time between sowing and reaping was for the most part filled with activities like weeding, doing house repairs, slaughtering pigs, celebrating special occasions, and, of course, praying for rain or sun. At harvest time the whole family would get underway to the fields at dawn. Men would harvest with a scythe onto which a cradle was attached to catch the grain and lay it down in even rows. (I remember well the general excitement when the scythe was replaced by a simple horse-drawn reaper.) The women would follow the men, gather the wheat, and bind it into sheaves. Toward evening the family would go home, where more work awaited them: weaving ropes for binding the sheaves, pounding the dulled scythe, and attending the cattle. After threshing (which was done using a single stationary threshing machine powered by a tractor for the whole village) the grain would be partly stored and partly sold so that other goods—like sugar, salt, and clothes—could be bought. It took a lot of hard work by the whole family to scrape a living.

Agribusinesses

Now, as I look through the window of my apartment (not too far from Sirač), I see a quite different way of work in the distant fields. In the sowing season I see (and hear) large tractors plowing deep to reach the fertile soil, then spreading evenly the carefully chosen seeds and fertilizing the fields with synthetic fertilizers. Small planes fly swiftly overhead spraying pesticides provided by the chemical industry. Irrigation systems make sure that moisture is never lacking in the soil. At harvest time large red combines spit threshed grain into the trucks, leaving neatly bound bundles of straw behind them. The agronomists with Ph.D.s from a nearby Institute for Agriculture are applying genetics to create plants that give much larger yields because they are better adapted to the soil conditions and resistant to insect attack and plant disease. Leaving aside for a moment the ecological damage done to the soil, the result is higher productivity and an increase in the total production of agricultural goods.

It is not easy to predict future developments in agricultural production, especially in view of the rapid advances of genetic engineering. We should at any rate expect further development of machines and automated systems "for harvesting [produce that is presently] still picked by hand, for transplanting plants, for adjusting the flow of irrigation water to individual fields."[10] The ubiquitous information technology will increasingly be applied to agricultural production, which will lead to "more efficient management of machines and energy" and will "help in other farm operations such as cost accounting, mixing feed and deploying fertilizers and other resources efficiently."[11]

Small-scale family farming and "agribusinesses" are two different worlds of work: the one is a world of production for one's own needs, of labor by one's hands, a world of the accumulated wisdom of generations, and dependence on the moods of nature; the other is a world of mass production, the world of technology and science that put nature under human control. Even though family farming is dominant in some societies, and in others still exists side by side with agribusinesses, it is a type of work that belongs to the past (except as a very deliberately adopted way of life, which is a different matter). It is only a matter of time before family farming will give way to one or another form of modern farming, making agricultural work around the globe an employment of a minority, as it is now in economically developed countries.

Manufacturing Production

Manufacture has gone through even more radical changes than agriculture. Looking at economically developed countries, one can distinguish three basic *stages* in the development of manufacture, which transformed the worker "from a craftsman to a machine operator to a machine overseer."[12] From a global perspective these three stages are three *types* of production that exist simultaneously in today's world; craftspeople rapidly giving way to machine over-

Manual Production

In the ancient world, the production of goods was done by skilled craftspeople. Describing crafts production in the Persian Empire, Xenophon says, "in small towns the same workman makes chairs and doors and plows and tables, and often this same artisan builds houses."[13] (Since in the Palestine of Jesus' time we find a similar absence of advanced specialization, Xenophon's description might also fit well the trade Jesus learned in his father's carpentry shop.)

The ancients clearly saw the advantages of increased specialization. Plato grasped the many benefits of the division of labor very well when he said that "all things are produced more plentifully and easily and of a better quality when one man does one thing which is natural to him and does it at the right time, and leaves other things."[14] Thus when socioeconomic conditions allowed for greater specialization, people took advantage of it. Xenophon describes as follows the artisan's work in Persian cities: "In large cities . . . , inasmuch as many people have demands to make upon each branch of industry, one trade alone, and very often even less than a whole trade, is enough to support a man: one man, for instance makes shoes for men, and another for women; and there are places even where one man earns a living by only stitching shoes, another by cutting them out, another by sewing the uppers together, while there is another who performs none of these operations but only assembles the parts."[15] But in spite of the increasing specialization of occupations (or, as the case may be, partly because of it), in the manual production "the workman continued to control his own working action. He modified the form of his product according to his own skill, taste, and judgment, the results of which were visible to him in the finished job. Even where a craftsman did not himself make the entire product . . . , he could clearly see the effect of his own handiwork."[16]

Machine Production

The nature of work changed radically with the introduction of machines. A more-or-less skilled craftsperson was replaced by a *machine operator*. Andrew Ure, an influential economist in the nineteenth century, defined the factory system of machine production in his time in the following way: "The factory system designated the combined operations of many orders of working people, adult and young, in tending with assiduous skill a series of productive machines, continuously impelled by a central power."[17] The key aspects of Ure's description of machine production are still true in many cases today (although not all aspects of his description apply to all types of machine production). Taking my cue from Ure's description of machine production (while modifying it in one respect), I will discuss some important aspects of manufacturing in the industrial era.

First, machine production was characterized by the "*combined operations* of many orders of working people." With the introduction of machine production the process of increased division of labor and fragmentation of work sets in and reaches its pinnacle in the implementation of "scientific management," according to which a worker should be performing her well-defined task on only a small fraction of a product.[18] Increase in the division of labor multiplies the number of workers whose cooperation is needed for production of a particular commodity. The functions that were formerly done by an individual worker now require the activity of a "combined worker."[19]

Second, the "*de-skilling*" of the worker characterized machine production. Ure was mistaken when he spoke of the "assiduous skill" needed for machine production. In many cases machine production demanded only minimally skilled workers.[20] Increased division of labor seems in fact to have decreased the skills of the laborers. As Adam Smith critically observed, the worker can be "as stupid and ignorant as is possible for a human creature to become" and still successfully perform "a few simple operations" required by the process of production.[21]

Third, workers in the "factory system" can be both "adult and *young*." Since machine production reduced work to a few simple operations, experience was just as dispensable as skill. Thus not only did the work of older men decline in value, also the greater "manual dexterity" of less-experienced children (and women) could be used effectively in production. An additional benefit was that women and children required less pay.

Fourth, the typical work of the industrial era consisted of "*tending* . . . a series of productive machines." There is an important difference between a simple tool and a machine. A tool transmits the activity of the worker to an object. The worker remains the primary agent of production, the tool being a means she is using. In machine production, on the other hand, the machine is the primary agent of production. The worker (as distinct from the owner of the machine) does not use the machine in the strict sense but "tends" it by "making adjustments for errors in the machinery, feeding it with materials, and checking output to make certain it is performing satisfactorily."[22]

Fifth, machines are "*continuously impelled* by a central power." Whereas a craftsperson could work at his own pace, the continuous operation of machines placed the machine operator under strict discipline. The "listless and restive habits" (Ure) of human beings (which were a constant nuisance to overseers, and a financial loss to employers) were effectively subdued by the mechanically controlled work-pace (though the ingenuity of workers in finding ways to break up the monotony of machine production cannot be underestimated[23]). The assembly line exacted especially strict discipline: each simple action of a particular worker had to be done in the allotted time, otherwise the flow of production would be interrupted.

In the machine production of the industrial era the worker did not control her own working action. She worked for the most part with mechanically enforced, monotonous regularity. Work became fragmentized into minute operations and the worker was robbed of understanding how her work related to the finished product. Such work was indeed a debasement of human nature.[24] But it did contribute to the creation of great economic wealth and prepared the way for the next stage of production, characteristic of the information era.

Automated Production

In economically highly developed countries the industrial era is giving way to the information era. At the root of this transformation, which might be the most momentous of all technological transformations so far in human history, lies the computer chip. In the words of its co-inventor, Robert Noyce, the chip represents a "true revolution: a qualitative change in technology . . . (that) has given rise to a qualitative change in human capabilities."[25]

As information technology is applied to industrial production, a new kind of factory is emerging in which "a hierarchical system of networked computers and microprocessors is used in the control and delivery of raw stock and parts in process, forming processes for metal (and other materials), production planning scheduling, mechanical assembly, inspection and testing, maintenance and repair, and accounting operations."[26] With the application of new technology, the industrial corporation, which was formerly dominated by manufacturing work, is being transformed into an information organization.[27] According to some scientists, by the year 2025 in the United States, for instance, all manufacturing will be done by robots.[28]

Information technology is changing the nature of human work in industrial production. The machine operator has become the *machine overseer*. Whether this change is making work more humane is another question. At best the worker is required to use her intellect in a creative way. "Each worker has to be pretty much in control of the process, has to understand it, know how to program the machines he is responsible for and reset them."[29] At worst the worker's "responsibility is limited to watching the gauges and lights, listening for the alarm bells, or reading the printout telling whether the production elements are functioning as programmed."[30] Whichever of the two scenarios actually takes place, yesterday's blue-collar worker will be replaced by an (either content or alienated) "information worker."[31]

A consequence of the application of information technology to industry is the increase in productivity, which in turn causes a radical reduction in the numbers of people employed in industrial production (which does not necessarily mean high unemployment rates). Since those who remain in industrial production will become "information workers," and since a very small percentage of people will be involved in agricultural production, in the future only a small minority of people will work on "nature," either in its fabricated or unfabricated form. Human work is ceasing to be material activity and is increasingly becoming mental activity. This is a radical change from past centuries, in which the predominant form of work was physical labor.

Two additional features characterize work in information societies. First, the growth of the service sector and the increasing necessity of teamwork make work in information societies more and more an interpersonal activity. In individualistic and egalitarian societies, as these are, the interrelations of human beings at work are not personal, however. They are best described as "contractual intimacy and procedural cooperation."[32] Second, whereas in traditional societies every "individual has a given role and status within a well-defined and highly determinate system of roles and statuses,"[33] modern information societies are characterized by a high degree of vocational mobility. At any given time a person will have several work roles, and these are likely to change in the course of her life. These are pluralistic societies based on plural activities of their members.

CRISIS OF WORK

Today we can observe a general crisis of work. It frequently surfaces in the negative attitude of workers toward their work. Many people are deeply dissatisfied with the kind of work they are doing. As seriously as we ought to take people's attitudes toward their work, they are not the most important aspect of the present crisis of work. As subjective states, attitudes are often deceptive. For various reasons, people often invest the most demeaning work with great significance.[34] On the other hand, they often superimpose on their most creative tasks dissatisfaction with other areas of their lives. More serious than subjective negative feelings about work is the *objective* crisis of work. For this reason in the following I will deal less with workers' subjective attitudes than with their objective situation as workers.

Various aspects of the crisis of work constitute problems for which a theology of work must propose solutions. As I develop a theology of work in the subsequent chapters, however, I will not come back to deal with each aspect of the crisis analyzed below. Such an endeavor would go beyond the scope of this study. But the theology of work I am proposing is consciously developed to provide a theological framework within which responsible and creative thinking about solutions to these problems is both required and possible.

Aspects of the Crisis of Work

In this section I will analyze only those aspects of the crisis of work I consider particularly important. The crisis can be felt in various degrees throughout the world. But the Third World is suffering under it disproportionately more than the First World or even the (disapearing?) Second World. This is particularly the case with respect to the aspect of the crisis of work that I will analyze first: child labor.

Child Labor

While paying a visit to some people who work at the recycling plant at the garbage dump near Osijek, the city where I live, I met a young boy. He is nine years of age, an abandoned child in the care of distant relatives. After twice failing first grade, he left school illiterate. A shack at the garbage dump became his home in the summer and for a good part of the winter, too. He works from sunrise to sunset, trying to survive, often by eating the food he finds in the garbage.

In economically highly developed countries (and in my home country, Yugoslavia, as well) such tragedies happen on the fringes of society, if at all. We associate them with the beginnings of the Industrial Revolution there. At the end of the eighteenth and the beginning of the nineteenth centuries, it was the plight of small children who had to work up to eighteen hours a day that pricked the consciences of Western Europeans. Today, when children work in these countries, however, they work, not to survive, but to earn spending money to acquire the latest line of fashion clothing or some high-tech product. Neither their circumstances nor their parents force them to work. Instead, the parents' concern is "to fill those little sponges with knowledge as early as possible," creating "superbabies" and "whiz kids."

In many parts of the world, however, child labor is not an exception, but a rule enforced by a dire need. It is estimated that 50 to 200 million children under age fifteen are in the world's work force. Many more (40 million street children in Latin America alone) work without being gainfully employed, performing various activities that "contribute to the maintenance of the household and to the well-being of its members."[35] Often children feed the families because the mothers have to take care of their smaller siblings and the fathers are unemployed.[36] Children are often as productive as adults, but for a fraction of the cost. And employers have less trouble putting them up to dangerous and crippling jobs. They are among the most exploited and physically and psychologically abused workers in the world's work force. The majority of working children are "condemned to a cruel present and to a bleak future."[37]

Unemployment

In the world today, some 500 million people are either unemployed or grossly underemployed. The majority are those who have not yet worked (the young) and those who are too difficult to retrain (the old). With the increase of productivity due to rapid technological advances and with the expected high growth of the world population, in many parts of the world the future does not hold good prospects for employment. Some researchers estimate that by the year 2000, "the world's job-seekers are expected to increase by another 750 million."[38] Particularly hard hit will be economically developing countries, not only because of their higher unemployment rates but also because they have comparably fewer resources available to create employment and mitigate the personal and social costs of unemployment than economically developed countries do.

Among the young unemployed, many *handicapped* are caught in a hopeless situation (in particular in poorer countries where their concentration is proportionately higher than in wealthier countries). Their handicap gives them a double disadvantage. First, since they impede the efficiency of the normal educational process, they are placed in specialized and often inferior educational institutions that prepare them inadequately for potential future jobs. Second, even when they are highly competent for a particular job, they have comparatively more difficulty getting employed than those who are not handicapped.[39] In many countries the number of unemployed handicapped is growing faster than the number of unemployed in general.[40]

There are severe *personal costs* of unemployment. Most people need to work in order to meet their basic needs. Being laid off makes a serious dent in the budget of the unemployed in wealthier countries, and in many cases this means dependence on welfare. In economically underdeveloped countries, unemployment means abject poverty, bad health conditions, hunger, and a decreased life span. Furthermore, for people in many diverse cultures, work contributes significantly to human identity and defines a person's place in society. A young woman from Belgrade, Yugoslavia, who searched in vain for a job, described her experience of unemployment as follows: "I am alienated. When you don't work you are not a member of society and you feel superfluous. It is different when you are employed: you speak differently and you breathe differently. It doesn't matter how high the salary is. Those who don't have a job don't fit anywhere." This young woman experiences the pain of unemployment not so much through financial loss as through the psychological damage of losing her sense of identity as a valuable and appreciated member of society.

There are also high *social costs* of unemployment. The unemployment of every person denies society the creative contribution that person can make by employing his gifts and talents. At the same time, society has to set aside a sizable amount of resources to assist the unemployed and deal with the rise in crime and drug abuse associated with joblessness.

Discrimination

To illustrate the problem of discrimination in work, I will consider gender discrimination (although the problem of racial discrimination could serve just as well). After industrialization the world of work (or the world of what came to be considered work[41]) became predominantly a male world. In recent years, however, especially in developed countries, women have been flooding into the job market. But despite the significant progress that has been made toward gender equality in many countries, sex discrimination in work continues to be a problem for many women. (And as rigid gender roles are breaking down, discrimination is becoming increasingly a problem for men, too![42]) I will not discuss the reasons for discrimination against women here (except to say that it need not necessarily lie in a conscious attitude of male prejudice or malice).[43] It is well documented that women are "either pushed or pulled into a narrow range of occupations"[44] that are underpaid, offer little status or security, and few promotional opportunities and fringe benefits. To illustrate the underpayment of women's work from a country in which feminist efforts are most vigorous: in the U.S., women still earn about 60 cents for every dollar that a male with a comparable education earns (projections indicate that by the year 2000, they will still earn only 74 cents).[45] To mention another aspect of discrimination, studies have shown that in areas of work in which both men and women are available to staff a role, men and women are usually assigned different job titles, with men getting the better title. Studies show that such patterns of segregation "are difficult to reconcile with the perspective stressing employer rationality on the one hand or worker's vocational 'investments' on the other hand. Accordingly," Bielby and Baron continue, "we suggest that sex segregation is built into the hierarchy of organizational positions and is sustained by sex stereotypes and workplace social relations."[46]

Dehumanization

When a reporter visited the garbage dump near my hometown, a woman working there told him: "You can talk with us, but don't take a picture of me. I don't want my husband to find out from the newspaper where I work." She felt ashamed of her work. It was dehumanizing because it was below her *dignity* as a human being.

Work is not dehumanizing only for those who have to deal with waste products thrown away by consumers. Though it might strike us as less abhorrent than rummaging through the garbage the whole day, work in the production of consumer goods is often no less dehumanizing. Think of work on an assembly line: stupefyingly simple actions done with monotonous regularity. Arthur Hailey, has described it well in his novel *Wheels:*

> He was learning: first, the pace of the line was faster than it seemed; second, even more compelling than the speed was its relentlessness. The line came on, unceasing, unyielding, impervious to human weakness or appeal. It was like a tide which nothing stopped except a half-hour lunch break, the end of a shift, or sabotage.[47]

True, at the peak years of assembly-line production, only 5 to 6 percent of the work force in industrialized nations worked on the assembly line.[48] Yet it must be a cause of concern when even a minority of workers is forced to perform, day-in, day-out, a few simple operations. Such work is an assault on human *creativity* and is dehumanizing.

There is another important aspect of dehumanization in work. In many countries (especially in the Third World), work laws are either nonexistent or not enforced. This gives the management a free hand to behave as tyrannical monarchs "whose whole time is occupied in contriving how to get the greatest quantity of work turned out with the least expense."[49] Since workers often desperately need jobs to keep families alive, they are powerless to fight the despotic structures in which they are used and abused. Their work is demeaning to the human spirit because it deprives them of *freedom*. Together with the denial of human dignity and creativity, being deprived of freedom in work is one of the most important aspects of the dehumanization of work.

Exploitation

Like the forced labor of children, exploitation of employees in economically developed countries seems by and large a matter of the dark past (although there are still discriminatory practices, particularly along the lines of race and gender, that result in exploitation). In many economically developing countries, however, workers are still being exploited. Within one country there are rich elites that have amassed material resources (land in particular) through morally (though not always legally) flawed ways and left the masses without resources to work with. The destitute majority is at the mercy of a wealthy minority, which dictates the unjust price of labor in order to increase their wealth at the expense of the poor.

The population of economically developing countries is also exploited by economically developed countries (often in alliance with the wealthy elites in exploited countries). It is notoriously difficult to determine the causes of the great disparity of wealth and income in the modern world.[50] It is too simple (and in many cases untrue), for instance, to ascribe the disparity in wealth distribution to the "existence of wicked capitalists deliberately setting out to grind the face of the poor."[51] The inequality is often a result of the internal cultural, political, or economic problems of the poor countries themselves (such as mismanagement, extravagance, corruption, lack of self-discipline in work, fatalism, or high population growth).[52] But frequently the prices at which developing countries must export their commodities in order to import food and manufactured goods are unfairly determined by powerful developed countries. It can also be persuasively argued that transnational corporations that employ workers and use the raw materials of a country while taking by far the largest share of the profits are exploiting the host country, in spite of bringing skills and capital into the country. At any rate, there is no doubt that the rich North (one-fourth of the world population, which receives four-fifths of world income) often uses its power to write the "rules of the game" and "determine North–South economic relations to its own advantage"[53]—especially in relation to smaller countries.

Ecological Crisis

The level of public interest in the environmental issues fluctuates over the years. In the eighties they got significantly less public attention than in the early seventies, while in the early nineties they seem to be again at the center of worldwide attention (especially in Eastern Europe, after the lifting of the Iron Curtain has made manifest the catastrophic extent of the damage). In any case, lack of concern for ecological problems certainly had more to do with the information saturation of the public, their retreat into the private sphere, and the "forced choice in the closed system" by the media than with any substantial advances toward a solution of ecological problems. Air and water pollution, diminishing natural resources, population increase, the destruction of wildlife and wilderness, soil erosion and the growth of deserts, as well as the endangering of our planet's life-support systems have not receded over the past twenty years, to say the least. Even while I have been writing this book, two well-publicized ecological disasters have taken place in Europe. The meltdown in the Chernobyl nuclear plant left our part of the world (Yugoslavia), which was by no means hit the hardest, without uncontaminated fresh food for some months (not to speak of the illness, genetic deformation and death caused to animal and human life in the more immediate vicinity of the catastrophe). The dumping of toxic waste in the Rhine by a Swiss pharmaceutical plant almost destroyed all life in that river. Even more disturbing than such ecological holocausts is the flippant attitude of many Third World leaders who regard ecological concerns as a luxury (or a trap set by the competitive North) that they can ill afford. The result is heavy pollution in many parts of the Third World, especially in the big cities.

There is no need at this point to enter into the complex debate on the causes of ecological problems. It will suffice to note that all explanations of the ecological crisis (even the population explosion theory) intersect at the point where *human beings intervene in nature through their work*. By definition, ecological problems are problems arising "as a practical consequence of man's dealings with nature"[54]—"nature" being understood as the nonhuman environment in which human beings live. It is safe to claim that human work is *the* cause of ecological problems.

If the ecological crisis is a crisis of the whole life-system (as it certainly is), then the ecological crisis calls into question the kind of work that is causing that crisis. Such work is a form of long-term self-destructive behavior of the human race. Both ecological alarmists and optimists must agree that the quality of life for the human race, and indeed its future, depend on the capacity of human beings to learn how to work in a way that is cooperative with, and not destructive of, their nonhuman environment.

Causes of the Crisis of Work

I do not intend at this point to analyze in detail the causes of the present crisis of work, but to indicate at what levels we should look for those causes. We should distinguish between three levels of causes: personal, structural, and technological. For analytical purposes, I am treating the three levels separately, but in actuality they are interrelated. In fact, the technological causes of the crisis of work do not seem ultimately to form an independent category of causes. They can be explained either through personal or structural causes of the crisis. Yet, because of their profound influence on the world of work they deserve to be mentioned separately.

The personal and structural causes are *mutually interdependent*. But they are clearly *distinct*. The personal causes of the crisis of work influence the social causes either positively or negatively, and vice versa, but one set of causes cannot be taken as a sufficient explanation of the other. Both the typical liberal view (which traces all evil and good in a society to the decisions of individuals) and the typical socialist view (which explains all social problems by the existence of evil structures and expects radical improvement from their change) are inadequate, though not so much in what they affirm as in what they deny. In countries that have undergone a socialist revolution, bitter experiences from the past have brought about a growing recognition that the structural change of a society must be accompanied by the "moral change" of individuals (although the desired moral change would be more easily realizable if some additional structural changes occurred). On the other hand, there is also a recognition in capitalist countries that structural modifications of one kind or another are necessary for the well-being of a society.

Personal Causes

To a large extent different aspects of the crisis of work are rooted in the personal attitudes and actions of the people involved; for example, the corruption of higher management and government officials. This constitutes a serious problem worldwide, but it is felt most acutely in some economically developing countries where it contributes a great deal to the pauperization of the people. Though some types of socioeconomic structures and legislation may be more conducive to corruption than others, it seems that no amount of structural and legislative change can completely eradicate it. Speaking of the need for honesty in the workplace, Ruskin rightly claimed: "Get that, you get all; without that, your suffrages, your reforms . . . are all in vain."[55] Corruption is fundamentally a problem of individual integrity. The solution lies in personal transformation.

Another example of the way management contributes to the crisis of work is the mistreatment of particularly vulnerable workers. This happens even in countries where legislation carefully guards the rights of employees. In his controversial book *Ganz Unten*, Günther Walraff exposed the unscrupulousness of some German employers' dealings with migrant workers. To make a buck, employers were willing to send unsuspecting workers to perform cleanup jobs in nuclear plants, even when the employers knew that this would have fatal consequences for their workers' health.[56]

But there are also personal problems of *employees*, that contribute to the crisis of work. In addressing the issue one runs the risk of blaming the victim. But even if one shares the preferential option for the poor (as I tend to do), one needs to recognize that some people are unemployed because (for instance) they do not want to work. And alcoholism, which is often portrayed as a consequence of being laid off, frequently in fact causes unemployment. To take another example, as consumers, employees contribute to ecological problems through their unwillingness to give up certain conveniences. True, industrial propaganda significantly influences ecologically irresponsible personal choices. But consumers cannot be exonerated by casting the blame completely on industrial propaganda.

Structural Causes

Some causes of the present crisis of work are built into the *structures* of economic life. The causes of unemployment, for example, that are found in the behavior of unemployed individuals seem minor compared with the causes that are found in the conditions under and rules by which the "economic game" is played. If the point of the economic game, which all its rules should be subservient to, is to maximize profit, then a corporation is likely, for instance, to transfer capital from the economically developed home country to an economically underdeveloped host country. If there is no one to mitigate the consequences, such transfers result in much suffering: they are likely to contribute to unemployment in the home country and can result in dehumanizing work in the host country (though they might also bring much-needed work to the country).

I take it for granted that individual employers and employees are not, strictly speaking, determined by the structures of economic life. If they are inspired by higher causes and are willing to take the punishment in economic terms, they can break the rules of the game. But the question is not how they *can* behave, but how they are *likely to behave* and how in fact they *do behave* within a particular economic structure. As a rule, employers shun economic punishment and play according to the rules of the game. Economic institutions thus often "place narrow limits on the choices of the best men who work within them."[57]

Technological Causes

Technological development is an ambiguous process with respect to its effects on workers. Technological innovations have clearly benefited workers, yet they have also contributed significantly to the present crisis of work. Machine production not only resulted in increased economic growth and a better standard of living for whole populations, but it also produced the degradation of the worker. A skilled craftsperson who controlled her work and saw its results in the finished product was turned into a mindless machine, mechanically doing a few simple operations. More recent information technology has relieved many blue-collar workers of some dangerous and monotonous jobs, but at the price of significantly de-skilling other jobs. Technological development has also significantly damaged the natural environment.

To solve the problem of the technological causes of the crisis of work, we ought not give up technology and return to preindustrial modes of production. In the initial stages of the Industrial Revolution, both in England and on the Continent, workers raided factories and smashed the machines, thinking that these were depriving them of their livelihood. Today some people look back nostalgically at the days of primitive technology against which their fellows were rebelling two centuries ago and suggest a moratorium on the implementation of new information technology in industrial production. The problem is not new technology itself, however, but the particular way the technology has been made and put to use.[58] The solution must therefore lie, not in the abolition of technology, but in its proper construction and use.

CHAPTER 2

Dominant Understandings of Work

A theology of work that does not want to miss the mark will consider, not only the character of work and its role in people's individual and communal lives (Chapter 1), but also the way people have come to *understand* their work. Theories of work are part and parcel of the reality of work that a theology of work needs to reflect on. For a theology that wants to be relevant to the contemporary economically developed and developing societies, there is no getting around an analysis of the dominant philosophies of work in these societies. Thus, in this chapter I will discuss Adam Smith's and Karl Marx's theories of work.

Dealing with thinkers who lived in the eighteenth and nineteenth centuries might seem an odd way to analyze the dominant contemporary understandings of work. Yet Marx and Smith influenced the reality and understanding of work in today's world more than any other thinkers, past or present. As is well known, the two philosophers and economists—as both Smith and Marx were—are the progenitors of modern capitalism and socialism, which in their various versions still dominate both economic life and thought in the world today.

Of course, modern capitalist and socialist economies are not run exactly according to the principles and rules laid down in the *Wealth of Nations* and *Das Kapital*. Over the years economists in both capitalist and socialist worlds have significantly modified the theories of Smith and Marx. These modifications are nevertheless not a kind that would prevent us from gaining insight into the dominant contemporary understandings of work by analyzing Smith's and Marx's concepts of work. First, the modifications often amounted to the incorporation of elements of one economic system into the other. As a result there are no pure market or pure planned economies. Each economic system combines to greater or lesser extents the elements of both, the difference between them consisting in the predominance of elements from one or the other model. Second, later modifications often are refinements of particular aspects of Smith's and Marx's economic theory, leaving the main features of the underlying social philosophies and theories of work untouched. These have not been significantly altered, even when modifications include paradigmatic shifts in understanding of the operation of economy—such as when "market" is no longer conceived as analogous to "machine" (Smith), but as analogous to "biological system" (Spenser) or "game" (Hayek, Friedman).[1]

In analyzing Smith's and Marx's theories of work, one may be tempted to give these thinkers labels that fit one's ideological preferences. It is, however, more profitable to seek to understand them and learn either from their grave (but rarely fatuous) mistakes or from their valuable insights. Smith is not a moral cripple who would hear of nothing but ruthless self-interest. Those who interpret him in this way had better read, not only his important work in moral philosophy, *The Theory of Moral Sentiments*, with which he identified his entire life, but also the passages from the *Wealth of Nations* in which he criticized the ravaging consequences for human nature of the division of labor. And Marx is not a red devil who had no respect for individual freedom and creativity and desired only to dispossess the industrious and wealthy. That freedom and creativity in all spheres of life, particularly in daily work, were the most important goals of his scholarly pursuit can be seen both from his early works (e.g., *Economic and Philosophical Manuscripts*) and his late works (e.g., *Grundrisse*).[2]

ADAM SMITH'S UNDERSTANDING OF WORK

The very first sentence of Smith's seminal work *Wealth of Nations* testifies to the importance he ascribed to human work: "The annual labour of every nation is the fund which originally supplies it with all the necessaries and conveniences of life."[3] It is one of Smith's most significant contributions to the development of economic thought that he singled out human work as virtually the only source of economic wealth and placed it at the center of economic theory.[4]

But for Smith, work was not only the main source of economic wealth. It also provided the structure for the whole fabric of society. He was the first thinker who proposed what was to become a very influential thesis: that historical forms of sociopolitical structure and of intellectual superstructure are determined by the "modes of production" predominant in a given period.[5] "The four stages of society," wrote Smith, "are hunting, pasturage, farming and commerce."[6] This four-stage theory of social development provided the structural outline that enabled him to trace "the gradual progress of jurisprudence, both public and private, from the rudest to the most refined ages, and to *point out the effects of those arts which contribute to the subsistence, and to the accumulation of property, in producing corresponding improvements or alterations in law and government.*"[7] Though he did not consider the economic factor to be the sole explanation of social, political, and intellectual history, as Marx later seemingly did, in Smith's view the economic factor was the dominant one.[8]

Material production was for Smith the kind of work that was economically and socially important. Drawing on Locke's reflections on the relation between work and property,[9] Smith maintained that labor "that fixes and realises itself in some particular subject or vendible commodity, which lasts for some time at least after that labour is past"[10] is productive and hence the most important form of labor. The work of a pastor, philosopher, or politician is unproductive because it "perish[es] in the very instance of [its] performance."[11]

Smith turned on its head the traditional relationship between economic activity and spiritual, intellectual, and political pursuits. For Aristotle, for instance, the goal of political activity was to ensure the good life of the citizens, and he considered it the highest form of practical activity.[12] Smith, on the other hand, considered political and intellectual activity subservient to economic activity and maintained that economic activity makes possible the good life in that it creates wealth and fosters civilization. Given wings by constantly increasing productivity, Smith's stress on material production made history. In the modern world, participation in material production—whether by manual or intellectual work—has become the most significant form of work, due to the prevailing persuasion that economic activity holds the key to human happiness.

Smith's stress on the importance of work for economic and civilizational progress (together with Fichte's and Hegel's more philosophical emphasis on activity) exerted a formative influence on Marx's understanding of the significance and function of work. Smith and Marx share the belief in the centrality of work in individual and social life. They differ in perspectives on the purpose of work, division of labor, and the resulting alienation of workers in capitalist societies. In order to set the stage for a later analysis of Marx's theory of work, I will continue the discussion of Smith's theory of work under these three headings.

Purpose of Work

In Smith's *Theory of Moral Sentiments*, we read: "Man was made for action and to promote by the exertion of his faculties such changes in the external circumstances both of himself and others, as may seem most favourable to the happiness of all."[13] At first glance the initial part of the sentence reads like a weighty pronouncement about the nature of human beings and the anthropological significance of work. As birds were made to fly, Smith seems to be saying, so human beings are made to work. Work is the great end for which human beings were created.

The second part of the sentence, however, makes it clear that it would be wrong to read the first part this way. For there we are told that human labor is not, as such, a purpose of human life but only a means for achieving the "happiness of all." Labor is not an essential characteristic of human beings without which they could not be human. It is merely a means to satisfy the "desire of bettering our condition"—a desire that, in Smith's view, is one of the distinguishing marks of human beings.

Since he sees work as a means of achieving happiness, it is natural for Smith to single out "consumption" as "the sole end and purpose of all production."[14] This is the purpose of the work of all the members of society—whether they do manual or intellectual work. Smith told his students: "All the arts, sciences, law and government, wisdom and even virtue itself[!] tend all to this one thing, the providing of meat, drink, payment and lodging for men."[15] People work, not because work is an expression of their humanity, but in order to satisfy their needs, ranging from physical necessities to fascination with "frivolous objects."[16] In spite of its centrality in Smith's economic theory, work does not have human dignity. It has only usefulness. If ever-expanding human needs that grow out of human desire for betterment could be satisfied in any other way, there would be no point in working.

Modern economic analysis, which takes work to be "a necessary evil in order to obtain purchasing power over goods and services,"[17] remains close to Smith's valuation of work. He considered work generally as something negative, something to be endured.[18] For when a person works, Smith thought, he always lays down a "portion of his ease, his liberty, and his happiness"[19]—a statement Marx liked to quote with no small abhorrence. The "sweets of labour," especially in the case of "inferior employments," as Smith calls them, "consist altogether in the recompense of labour."[20] Hence only necessity will force people to work.

Work being something negative, the ultimate interest of a person is not to work but to "live as much at his ease as he can." In fact, one of the purposes of acquiring wealth is to avoid working. Since he considered labor virtually the only source of wealth, it is natural that Smith would define wealth as "a certain command over all labour."[21] A person who commands the labor of others has the source of wealth under her control. Wealth is the power to dispose of the activities of others.[22] The more wealth one has, the more one will be able to "avoid irksome labour and impose it on others."[23]

Division of Labor and Pursuit of Self-Interest

A tension characterizes Smith's thinking about the value of human work. Individually, the goal of every person is to avoid working because anthropologically, work is something clearly negative. Societally, however, the goal is to increase the quantity and quality of work because economically, work is something thoroughly positive; it is the primary factor in economic growth on which the whole progress of society depends. The tension between the anthropologically negative and socially positive valuations of work explains the seemingly curious phenomenon that the idolizing of work coincided historically with the beginning of strenuous efforts to eliminate the human factor from production.[24] When one makes the observation, however, that in spite of the rhetoric, work itself was not idolized—it was viewed as a mere means—rather the effects it produced were, then the curious phenomenon becomes explicable. For efficiency of production increases with the elimination of the human factor.

Division of Labor

The most fundamental way to eliminate the human factor and increase efficiency is the division of labor. It lies at the root of all rationalization of labor and labor-saving technology. Smith was certainly not the first thinker to emphasize its importance. As I have pointed out earlier,[25] very early in the history of civilization people have discovered the benefits of the division of labor. Schumpeter was right to call it "an eternal commonplace in economics."[26] There are, however, three distinctive emphases in Smith's thinking on division of labor that have been very influential. One concerns the anthropological significance, the other two the economic and sociopolitical significance of the division of labor.

First, division of labor shapes human nature. From Plato on, it was common to base division of labor on natural differences between human beings. "There are diversities of natures among us which are adapted to different occupations," Plato claimed.[27] Smith saw things the other way around. While he was not blind to the innate differences between people, he considered the differences by and large not the cause but "the effect of the division of labour."[28] Whereas Smith denied that work belongs to the essence of human beings, he affirmed that it significantly shapes human nature. He anticipated the modern notion that it is not only who people are that determines what they do, but also what they do that determines who they are.

Underlying Smith's views on the anthropological significance of the division of labor is Lockean anthropology. In order to arrive at the conclusion that human beings are all equal "to each other," Locke started with the premise that human beings are "all the Workmanship of one Omnipotent, and infinitely wise Maker."[29] As created by God they are "furnished with like Faculties, sharing all in one community of Nature."[30] The obviously existing differences between human beings are not implanted by God and hence not inherent to human beings. They are "socially acquired by virtue of different economic positions," and these in turn result from the exercise of the individual's freedom of choice.[31] Hence, whatever the difference between human beings, that difference is the result of their own activity.

Second, with Smith, division of labor becomes practically the only factor in economic progress. It can "*alone* account for ... superior opulence which takes place in civilised societies"[32] from which, "in a well governed society,"[33] both rich and poor benefit, and is a presupposition for cultural development. Smith was the first person to put such a burden on division of labor. And he was persuaded that the division of labor could carry it. For it effects "the greatest improvement in the productive powers of labour"[34] by increasing the dexterity of every laborer, by saving time, and by stimulating inventiveness. As we will see shortly, Smith had a divided mind on the influence of the division of labor on the skills and inventiveness of workers. Undoubtedly, though, division of labor did increase productivity by making human work a mechanical and hence more efficient activity.

Third, division of labor was in Smith's view the principle that structured and sustained modern societies. Their members are highly interdependent because in "civilised societies," people need to cooperate and assist one another in order to survive and prosper. The ties between them rest neither on divine decree nor on the will of some charismatic leader. They are an outgrowth of their individual desire to better their standing and of the insight that work using the principle of division of labor is more efficient.[35]

Division of labor is advanced and the ties that bind people are strengthened by a "propensity in human nature . . . to truck, barter and exchange one thing for another."[36] Smith did not consider this propensity *the* distinctive characteristic of human beings, as is often claimed.[37] But he maintained that it is common to all people either as *an* original principle of human nature or, more likely, a necessary consequence of the faculties of reason and speech.[38] As rational creatures, human beings divide labor and exchange goods and services in order to better the condition of each individual.

Pursuit of Self-Interest

Division of labor is the main way individuals in "civilized societies" assist one another. Smith soberly explains their readiness to do so, not by their self-sacrificial love, but by their self-interest. It is most likely, Smith wrote describing general human experience, for a person to get help from others "if he can interest their self-love in his favor, and show them that it is to their own advantage to do for him what he requires of them. Whoever offers to another a bargain of any kind, proposes to do this." Smith continues with what has probably become his most frequently quoted statement: "It is not from the benevolence of the butcher, the brewer, or the baker, that we expect our dinner, but from their regard to their own interest."[39] Anyone who objects to this way of putting things is blind to the obvious; namely, that the pursuit of self-interest is in fact, as Smith put it, "the general principle which regulates the actions of every man."[40]

But Smith claimed not only that people do pursue their self-interests in interaction with one another but that they need do nothing more than that. This is not to say that he was a ruthless egoist. In *Theory of Moral Sentiments* he wrote that "to restrain our selfish, and to indulge our benevolent affections, constitutes the perfection of human nature."[41] Yet Smith maintained that in economic activities there was no need for the exercise of this perfection. The pursuit of self-interest sufficed. Smith shared the liberal belief in the harmony of self-interest with the good of civil society, which was maintained by the "invisible hand." Whether one takes it to be the hand of divine Providence or as the established natural order, the "invisible hand" transformed self-interested rational labor of individuals into a system of mutual service. There was no need for virtue to realize common good. It realized itself, so to say, behind the backs of industrious self-seekers.[42]

Alienation

Alienation is no marginal phenomenon in "civilised societies." It affects the majority of the population ("the *great* body of the people") and causes their *almost entire* corruption and degeneracy." Moreover, since Smith claimed that "the great body of the people *must necessarily*" suffer alienation, he obviously did not believe that alienation is accidental to the character of these societies but rather that it belonged to the inner logic of their functioning.[43] Alienation is the price these societies have to pay for being economically advanced and civilized.

Alienation expresses itself basically in three interrelated ways: workers are powerless, they are exploited, and they become estranged from themselves. First, in Smith's view the worker is effectively powerless "in disputes with employers; in determining the number of pieces in piece work, and the speed or length of time of his work in hourly labour; and powerless to decide the form of payment."[44] As Smith says in an early draft of *Wealth of Nations*,

> The poor labourer who . . . while he affords the materials for supplying the luxury of all the other members of the commonwealth, and bears, as it were, upon his shoulders the whole fabric of human society, seems himself to be pressed down below ground by the weight and to be buried out of sight in the lowest foundation of the building.[45]

Second, Smith speaks of the exploitation of the worker. He describes the existing conditions of workers as "oppressive inequality,"[46] which was at least partly due to exploitation. In any given society there is always a minority

> who don't labour at all, and who yet, either by violence, or by the orderly opposition of the law [!], employ a greater part of the labour of the society than [the majority]. The division of what remains too, after this enormous defalcation, is by no means made in proportion to the labour of each individual. On the contrary those who labour most get least.[47]

Third, workers are estranged from their true selves. Smith's views on the formative influence of work on human beings provide the backdrop for his critique of this form of alienation. "The man whose whole life is spent in performing a few simple operations . . . generally becomes as stupid and ignorant as is possible for a human creature to become."[48] This is especially distressing because the faculty of reason is the distinctive characteristic of human beings.

Since they live in "drowsy stupidity," they are, in Smith's view "mutilated and deformed in an . . . essential part of the character of human nature."[49] In fact, Smith claims that "all the nobler parts of human character may be, in a great measure, obliterated and extinguished" in the great majority of the people.

In Smith's writings we find no suggestions on how to overcome alienation. It is not that Smith thought that all work must be alienating in the sense described. He knew not only of "inferior employments" in which people suffered by the division of labor, but also of "superior employments" in which more desirable work was performed. He does not elaborate on these, but his notion of desirable work can be gleaned from his sporadic descriptions of people of higher rank who escape the alienating consequences of division of labor. Their work is characterized by complexity ("not simple and uniform") and its demand for creative thinking ("exercises the head"). Their workday is, moreover, shorter, so that they have "a good deal of leisure, during which they may perfect themselves in every branch of either useful or ornamental knowledge."[50]

It is significant to discover the features of "superior employments" also in Smith's descriptions of work in "barbarous societies" preceding the age of manufacture and foreign commerce.[51] Obviously, he believed that the alienating consequences of division of labor are a burden that must be endured because the division of labor is the key to the progress of civilization. Faced with the choice between the economic prosperity of society and the alienation of laborers, between opulence that extends itself "to the lowest member of the community" and the "corruption and degeneracy" of "all the inferior ranks of people"[52] Smith opted for economic progress and decided that we must put up with the division of labor.

KARL MARX'S UNDERSTANDING OF WORK

Faced with the alienating consequences of division of labor, Marx chose differently than Smith had: he decided that division of labor has to go, and with it the whole structure of a market economy. Underlying this decision are his views on the anthropological significance of work.

Theologians interested in Marx's thought tend to turn to his early writings. The study of Marx's later writings in which economic analysis prevails seems to bear fruit less palatable to them than the study of his philosophical and sociological thinking in the early writings. For two reasons I will not follow suit and will base my analysis of Marx's understanding of work on his later writings (primarily on his posthumously published *Grundrisse*, which contains the most comprehensive presentation of his theory in general and his understanding of work in particular).

First, in spite of the greater interest by philosophers and social scientists in the early Marx, the influence of the late Marx is incomparably greater. It is Marx's late writings that have decisively shaped existing socialist societies. Since my task here is to analyze dominant contemporary understandings of work, I must deal with Marx's later writings. Second, even in his later writings, Marx is led by the same humanistic framework and desire to overcome alienation in work which characterize his early writings. If anything, his thought has matured through his careful study of economy.

Marx devoted most of his attention to an analysis of capitalist production and critiqued its alienating character, but he also (though significantly more mutedly) praised its civilizing influence. I will discuss these two aspects of Marx's analysis of capitalist production in the second and third sections below. The last two sections will be devoted to his reflection about work and leisure in the communist society. I will start with his analysis of the nature and function of work in general independent of the socioeconomic context in which it is done.[53]

Nature and Purpose of Work

Although people have always worked and will continue working as long as they exist, in different periods of social history they have worked in different ways. All historically contrasting forms of work, however, have certain common features. In Marx's view these stem from the fact that work is a process between nature as an object and human beings as acting subjects.[54]

This way of describing the relationship between human work and nature is correct but could also be misleading. It puts a one-sided emphasis on the distinction between human beings and nature. For

Marx, however, work is no "supernatural creative power" that human beings exert over nature. It is a *natural* power, because human beings are fundamentally natural beings. The unerasable distinction between human subject and natural object notwithstanding, when human beings work on nature, nature, through them, works on itself.

As a process between human beings and nature, work has four features. Since human beings are rational beings, work is, first, a *purposeful activity*. What sets apart the worst architect from the best of bees is that "the architect builds the cell in his mind before he constructs it in wax."[55] The second essential element of work is the *object on which* work is performed. It can be provided either directly by nature (as for the most part in agriculture) or be "filtered through previous work" (as in manufacture).[56] The third element is the *objects with which* work is performed, or the tools. A tool is "a thing . . . which the worker interposes between himself and the object of his labour and which serves as a conductor, directing his activity onto that object."[57] Fourth, since human beings are essentially social beings, all work is an "appropriation of nature on the part of an individual within and through a specific form of society."[58] So work is a purposeful, social activity through which people, helped by tools, manipulate nature.

Marx states very clearly what the primary purpose of work is: it "mediates the metabolism between man and nature, and therefore human life itself."[59] Work is a means for human beings to keep their body and soul together. This seems to contradict some of Marx's earlier, most basic persuasions about work. In his early writings, work conceived of as a mere means is a feature of alienating work, not of human work in general.[60] But we need to keep in mind that for the late Marx, work remains *also an end* in itself—he never gave up the notion that human beings should enjoy work—and that for the early Marx, work was always *also a means* to maintain human existence.[61]

But for Marx work had a broader significance than simply to make human life possible. Self-realization and the creation of a human world through the humanization of nature are for Marx essential functions of work. As is well known, the early Marx stressed that human beings develop their physical and mental skills through nonalienating work. The same idea recurs in *Das Kapital*, where we read that a person acting on external nature "simultaneously changes his own nature," developing "the potentialities slumbering" in it.[62]

The idea that work serves to humanize nature is present in Marx's later writings,[63] but it figures especially prominently in *Economic and Philosophical Manuscripts*. There we read that, when worked upon, "nature appears as *his* [the worker's] work and reality."[64] Marx did not, of course, entertain the absurd idea that human beings, like God, create nature *ex nihilo*. Instead, he meant that through work they leave imprints of themselves on nature so that nature becomes, so to speak, their extended self. Through work they "reproduce" themselves in objective reality "and see their own reflection in the world that they have constructed."[65] Thus, although its primary purpose is to secure human existence, work functions at the same time as a means by which human beings develop their own potentials and create the world in their own image.

Alienation and Humanization of Work

The analysis of the capitalist society and of the alienation that human beings experience in it was the most important part of Marx's lifelong study of economy. The backdrop for his analysis is not nostalgia for some past mode of production destroyed by the coming of capitalism, but the expectation that a "higher mode of production" that is to evolve out of capitalism will overcome all alienation in work (and consequently alienation in all other aspects of human life).

Inversion of Means and Ends

One precondition of the capitalist mode of production is the advance of division of labor to the extent that the primary goal of production is not the use of products (either by producers themselves or their fellow human beings), but the exchange of products as commodities. A product that a worker makes has no value for her as a product, but only as a *means* to acquire another needed product or service. She is fully indifferent to what she is making. Marx contrasts this situation with a normative model in which workers produce for consumption. Correspondingly, in the communist society one will not produce in order to eke out a living or make profit, but in order to satisfy the concrete needs of one's fellow human beings.[66]

If products are only means, then the activity by which they are made, maintained Marx, "is not an end in itself . . . , but a means."[67] For two reasons work is alienating when it is only a means and not an end in itself. First, human beings can enjoy work only when they, at least partly, do it for its own sake. If work is a mere means then people do not work because they enjoy working but because they cannot survive without working. Work is then a form of forced labor. It is, as Smith maintained, a necessary sacrifice of their ease, liberty, and happiness. Marx was, of course, aware that, being dependent on nature, human beings have to work in order to survive. He considered work to be "an eternal necessity."[68] In that sense people are, no doubt, forced to work. But he believed that human beings should not work only because they have to work but also because they like to work. For it is in the nature of human beings that they have "a need for a normal portion of work,"[69] not just for the results of work.

Human beings have a need to work because work is their "life-expression." In fact, work constitutes their essence.[70] For, in Marx's view, the "whole character of a species" is determined by "the type of life activity," which for human beings consists in work as "free, conscious activity."[71] The central anthropological significance of work is the second reason why it becomes alienating if work is reduced to mere means. For then a person's very "*being* [becomes] only a means for his *existence*."[72] Marx expects that in the communist society, work will be a form of the "self-realization" of human beings and thus also a "free activity" that is an end in itself.[73]

As individuals work in capitalist societies, they pursue their own individual interests and those only. The common interest is recognized as relevant, but it is not the motive of individuals' actions. Instead, it is realized "behind the back of one individual's interest in opposition to that of the other."[74] Under such conditions, Marx claims, each person "serves the other in order to serve himself; each makes use of the other, reciprocally, as his means."[75]

Marx criticizes the activity of self-interested individuals (to which Smith ascribed a historically positive function of fostering public good) as debasement of human nature. The normative presupposition of his critique is the anthropological persuasion that what differentiates human beings from animals is not only that they work freely and purposefully but also that they work consciously for one another. They "relate to one another as human beings"[76] only when they "reach beyond" their particular need and are concerned about the well-being of others. Moreover, for individual and public interests to coincide it is not sufficient for each to pursue her own interest. All individuals must make the interests of others their own interest.

Division of Labor and Machine Production

I have already indicated that, in Marx's view, division of labor is alienating because, in the context of capitalism, it turns the worker, her activity, and her products into mere means. But advanced division of labor is alienating also because it changes the character of human work. The operations that were performed by an individual worker become severed and divided amongst a plurality of workers.[77] None of the structural features of the original integral work process gets lost, but they are, so to speak, spread out, and characterize the work of the "combined worker." In the process, the work of an individual worker "loses all the characteristics of art" and becomes increasingly "a purely mechanical activity, hence indifferent to its particular form."[78]

Marx criticized such mechanical activity on two counts. First, with Smith,[79] he objected that the division of labor crippled workers. Permanent mechanical repetition of a single operation has devastating consequences for the physical and mental health and development of laborers.[80] Second, the skills workers have lost through division of labor have been incorporated into a particular form of the organization of work. They are taken away from workers and come into the possession of the capitalists or managers who control their work. Thus division of labor "produces new conditions for the dominance of capital over labour."[81]

One of the positive aspects of capitalism is that, through "unlimited mania for wealth," it "incessantly whips onward" technological development.[82] By making use of the natural sciences, it transforms inherited tools and creates "the *machine*, or rather, *an automatic system of machinery*."[83] Machinery is not simply a more complicated tool. Tools serve "to transmit worker's activity to the object,"[84] while machinery, on the other hand, has a principle of action in itself, and a human person is there only to "transmit the machine's work" to the object by supervising it and guarding it against interruptions.

Marx was not critical of machinery as such. He expected that the communist society would be able to step up technological development. But he objected to the way machinery was used in capitalist production. For one, the worker no longer needs any skills to work. The machine itself is "a virtuoso, with a soul of its own in the mechanical laws acting through it." Human work can thus be reduced "to a mere abstraction of activity."[85] Second, workers lose their freedom. When they work with tools, it is they who use the tools. Now the machine uses them: their activity is "determined and regulated on all sides by the movement of the machinery, and not the opposite."[86] Third, as was the case with division of labor, the de-skilling of workers and the incorporation of their skills into machinery deepen the dominance of capitalists over workers because the capitalists control the machinery.

The normative presupposition both for Marx's critique of the division of labor and the capitalistic use of machinery resides in his anthropological persuasion that freedom and creativity are essential characteristics of human beings that need to be expressed also— indeed above all—in human work. Marx expects that in the communist society, workers will no longer be dominated by machinery because the machinery will belong to them and they will have sufficient technological know-how to understand how it functions. Thus they will no longer work as "drilled animals" but as free and creative agents. He even muses—quite unrealistically—that individual workers will possess the "accumulated knowledge of society" and that their work will become "experimental science, materially creative . . . science."[87]

"The Great Civilizing Influence of Capital"

In spite of his sharp criticism of capitalism, Marx was more subtle a thinker than to consider it simply a historical aberration. Because he, much like Smith, valued its "civilizing influence,"[88] he believed that capitalism is the necessary stage in the development of the human race toward the communist society. He particularly praised what he called the "revolutionizing" character of capitalism, its destruction of "traditional, confined, complacent, encrusted satisfaction of present needs, and reproduction of old ways of life," and its tearing down of "all the barriers which hem in the development of the forces of production, the expansion of needs, the all-sided development of production, and the exploitation and exchange of natural and mental forces."[89]

Using his own terminology, Marx mentions here three aspects of the civilizing influence of capital: technological advancement, enrichment of human personality through development of individuals' skills and needs, and the extension of human rule over nature. I already touched on Marx's appreciation of the technological development in capitalism and will not elaborate further on it. I need only mention that he considered technological advancement one of the primary preconditions for the remaining aspects of the civilizing influence of capitalism.

Dominion over Nature

An important aspect of the civilizing influence of capitalism concerns the human relation to nature. As the drive to make profit goads capitalists to constant technological advancement, so it also spurs them to scientific "exploration of all of nature in order to discover new, useful qualities in things."[90] The respect for nature that largely characterized all precapitalistic societies gives way to concern with the utility of nature. And powerful technology makes possible "the universal appropriation of nature . . . by the members of society."[91]

Marx praised the subjugation of nature as a civilizing achievement of capitalism. He never thought much of the romantic call to return to a life in unity with nature. Even "reactionary Christianity" was beyond such idolatry of nature.[92] True, in his early writings Marx was (less carefully reflecting and more) freely musing on the "genuine resolution of the conflict between man and nature" in communism.[93] But even there the conflict between human beings and nature was not to be resolved through a return of human beings to a peaceful unity with nature, but through the subjugation of nature by human work.[94] Capitalists knew how to subjugate nature, and Marx knew how to appreciate that. But he was persuaded that progress in the capitalistic manner of production is progress in the art, not only of robbing the laborer, but also of robbing the earth. By conquering nature, the capitalistic manner of production destroyed its ecological balance.

Since in communism people will not be blindly led by the profit motive and since the means of production will be in their possession, they will be able to plan production in a way that does not destroy nature. They will not only rule over nature but also over the possible negative consequences of that rule. Human beings are responsible for this because, in Marx's view, they are not the owners of nature but have only the right to use it with an obligation to leave nature "in a better condition to the coming generations."[95] Provided they do this, the task of human beings is to attain "the full development of the mastery . . . over the forces of nature, those of so-called nature as well as of humanity's own nature."[96]

Development of Human Abilities and Needs

The most important civilizing contribution of capitalism was that it produced the conditions for the development of "rich individuality."[97] Activity and enjoyment being the two basic aspects of human life, people with rich individuality will have both highly developed skills and highly developed needs. The one is the precondition for what Marx calls "all-sided production," and the other, for the enjoyment of "all-sided consumption."

Given Marx's criticism that work in capitalism cripples both the human spirit and body, it might seem strange to hear him speak of capitalism's contribution toward the development of the universally competent worker.[98] But for various reasons that are not completely persuasive, Marx seems to have believed that capitalism cannot continue to reduce work to performing a few simple operations. He predicted that modern machine production will compel capitalists

> to replace the detail-worker of to-day, crippled by life-long repetition of one and the same trivial operation, and thus reduced to a mere fragment of a man, by the fully developed individual, fit for a variety of labours, ready to face any change of production, and to whom the different social functions he performs, are but so many modes of giving free scope to his own natural and acquired powers.[99]

When this happens, then the "kingdom of freedom" will have come near.

In the communist society, all individuals will have all their potential fully developed. This will be possible because within the realm of material production (Marx calls this "the kingdom of necessity"), workers will not perform stupefying tasks but will work as creative scientists. Increased productivity resting on technological advances will reduce to a minimum even such necessary (but not alienating) activities. People will hence have a great deal of free time for their own artistic and scientific development.[100] This realm of free time is what Marx calls the "kingdom of freedom," and it represents the pinnacle of communism. The goal of a communistic society is to reduce the "kingdom of necessity" (without sacrificing universal wealth, which is a precondition of communism) and to spread the kingdom of freedom as the time for the "free development of individualities."[101]

One aspect of rich personality is the ability to perform with skill many complicated tasks. The other consists in being "all-sided in . . . consumption."[102] Leaning on Hegel,[103] Marx describes human nature as a "totality of needs and drives."[104] The more needs a person has and the more refined those needs are, the more developed the person is. Human beings differ from animals in that their needs are not stationary but dynamic; indeed, "limitless."[105] Even the simplest human needs (like the need for food) are not fixed. If not in their factuality then certainly in their character, these needs are a product of historical development.

Human needs develop by means of the invention of new products: the perception of new products creates a need for those products. Taking art as an example, Marx wrote: "The object of art—like every other product—creates a public which is sensitive to art and enjoys beauty."[106] Because of their incessant search for profit, capitalists are particularly inventive in creating new products that in turn give rise to new needs. More than any precapitalist economy, capitalism in this way creates the conditions for the universal development of human personality.

But only the *conditions*, not rich individuality itself, at least not for the vast majority of the population! Leaning on the wage theory of Ricardo, Marx maintained that the needs of a small, wealthy minority can develop in capitalism, but only because the needs of workers are at the same time being reduced to the need for the "barest and most miserable level of physical existence."[107]

Marx expected that, because of highly increased productivity, in communism all human beings would be wealthy. "All wellsprings of the societal wealth will flow abundantly," he predicted.[108] And if economic output grows constantly, then the needs of the whole population will be permanently developing.[109] Since in communism, as Marx envisioned it, people will have a great deal of leisure, they will also have time to enjoy the fruits of their work. In fact, in descriptions of the "kingdom of freedom," Marx consistently mentions "enjoyment" as one of its prime characteristics.

PART II

Toward a Pneumatological Theology of Work

CHAPTER 3

Toward a Theology of Work

Given the paramount importance of work in both liberal and socialist economic and social theory, it is remarkable that in our world dominated by work a serious crisis in work had to strike before church bodies paid much attention to the problem of human work.[1] Theologians are to blame for the former negligence. Amazingly little theological reflection has taken place in the past about an activity that takes up so much of our time. The number of pages theologians have devoted to the question of transubstantiation—which does or does not take place on Sunday—for instance, would, I suspect, far exceed the number of pages devoted to work that fills our lives Monday through Saturday. My point is not to belittle the importance of a correct understanding of the real Presence of Christ in the Lord's Supper but to stress that a proper perspective on human work is at least as important.

One might object that the most basic things in life are not necessarily the most important, and that it is hence superfluous to spend much time reflecting on them. Breathing is rather basic to life, but we do it twenty-four hours a day without giving it a second thought—until air pollution forces us to do so. Working, one might say, is much like breathing: its point is to keep us alive, and we need not bother with it until its function is hindered.

The parallel between breathing and working makes sense, however, only in a theology that subordinates the *vita activa* completely to the *vita contemplativa*.[2] As Thomas Aquinas' reflection on work illustrates, in such a theology the only real reason to work is to make the contemplation of God possible, first by providing "for the necessities of the present life"[3] without which contemplation could not take place, and second, by "quieting and directing the internal passions of the soul," without which human beings would not be "apt [enough] for contemplation."[4] But apart from the fact that work is necessary to provide for the necessities of the body and to quiet the passions of the soul, work is *detrimental* to human beings, for "it is impossible for one to be busy with external action and at the same time give oneself to Divine contemplation."[5] When a person inspired by the love of God does the will of God in the world, she *suffers* separation from the sweetness of Divine contemplation.[6] Where the *vita activa* is fully subservient to the *vita contemplativa*, there is no need to reflect extensively on human work, since, as a mere means to a much higher end, it is in the long run accidental to the real purpose of human life.

The complete subordination of *vita activa* to *vita contemplativa* that has been basic to much of Christian theology throughout the centuries betrays an illegitimate intrusion of Greek anthropology into Christian theology.[7] Faithfulness to our Judeo-Christian biblical roots demands that we abandon it. I am not suggesting that we should follow the modern inversion of the traditional order between *vita activa* and *vita contemplativa* and subordinate *vita contemplativa* completely to *vita activa*.[8] I am not even suggesting that we should place them on an equal footing. I do propose, however, that *we treat them as two basic, alternating aspects of the Christian life that may differ in importance but that cannot be reduced one to another, and that form an inseparable unity.*[9]

As soon as we ascribe inherent and not simply instrumental value to the *vita activa* (and thereby also to human work) we have answered the question of whether theological reflection on work is fundamental or marginal to the task of theology. Now another question faces us: What form should the necessary theological reflection on work take? Would an *ethic* of work suffice (as Christian theologians have thought through the centuries)? Or is a *theology* of work required? After arguing in the following section for a theology of work, I will deal with the way it should be crafted. I will end this chapter with a brief discussion of the formal characteristics a theology of work will have if developed in the framework of the concept of new creation.

A THEOLOGY OF WORK

Both the inherent importance of human work and the need to respond to the contemporary crisis in work in an emerging information society[10] call for the development of a *comprehensive contemporary theology of work*. The term "theology of work" is of recent date. According to M.-D. Chenu, who was one of the first to develop a theology of work, the term appeared first in the early 1950s.[11] It was introduced to express an important shift in the theological approach to the problem of work.

Work and Sanctification

Traditionally, the doctrine of *sanctification* has provided the context for theological reflection on the problem of work. This approach was introduced by the early church fathers, who developed some of the dominant features of the biblical understanding of work. In spite of treating the problem of work only as a subordinate theme, they provided the basic direction for most subsequent theological thinking about work. What we read in the writings of later theologians is for the most part variations on the church fathers' basic themes while they take into account a slightly changing historical situation.

Traditional Approach

Taking the doctrine of sanctification as their starting point, the early church fathers reflected on work from two main perspectives. First and foremost, they discussed *what influence the new life in Christ should have on a Christian's daily work*. Against the Greek philosophical depreciation of work, they affirmed that there is nothing disgraceful or demeaning about manual labor. Following the Old Testament, which portrays Adam in the Garden of Eden as working and caring for it (see Gen. 2:15), Clement of Alexandria, for instance, declared that it was "respectable for a man to draw water himself, and to cut billets of wood which he is to use."[12]

The early church fathers affirmed not only the nobility of work but also the obligation to work diligently and not be idle. Echoing the apostolic injunction to work with one's own hands (see 1 Thess. 4:11; 2 Thess. 3:10), they stressed that Christians should be "ever labouring at some good and divine work."[13] At the same time, they warned about the dangers of excessive work, admonishing their readers not to be "busy about many things, bending downwards and fettered in the toils of the world,"[14] but to take time for rest and worship. For Jesus Christ himself said that Mary, who sat at his feet and listened to his teaching, had chosen a better portion than her busy sister Martha (see Luke 10:38ff.). Early church fathers polemicized against human reliance on the results of work (wealth) as opposed to an attitude of dependence on God. They maintained that a Christian should not carry possessions "in his soul, nor bind and circumscribe his life with them."[15] From Jesus' story of the rich fool they learned that "a man's life does not consist in the abundance of his possessions" (Luke 12:15).

They also stressed that Christians should not work only to satisfy their own needs but also in order to have something to share with their needy fellow human beings (see Eph. 4:28). For they believed that those who are "without pity for the poor" and who are "working not for him who is oppressed with toil" followed the "way of the Black One."[16]

Early church fathers also insisted that there are occupations incompatible with Christians' new life in Christ, such as that of soldier. In disarming Peter (Matt. 26:52), the Lord "unbelted every soldier," making it clear that there can be "no agreement between the divine and human sacrament (i.e., military oath)." Hence a Christian may not make war, nor "serve even in peace."[17]

The second approach to work we find in the early church fathers is reflection on the *influence of work on Christian character*. Occasionally we encounter in their writings the quasi-heretical idea of the atoning function of work. For example, the Epistle of Barnabas affirms "working with thine hands for the ransom of thy sins."[18] But the more dominant (and orthodox) understanding of the influence of work on Christian character stressed that it served *ad corpus domandum*—to muzzle the evil and disobedient flesh. Especially later in monasticism, laziness was disparaged as "the enemy of the soul,"[19] and work, particularly if burdensome, was esteemed "as a spiritual exercise and discipline, a penitential practice."[20] For monks believed that work "allays concupiscence, forestalls temptation, and promotes humility and monastic equality."[21] Even Luther, who was no great friend of monasticism and its understanding of work,[22] valued the disciplining function of work and maintained that Christians are called not to idleness but "to work, against the passions."[23]

Reshaping Traditional Reflection

Reflection on work from the perspective of the doctrine of sanctification is indispensable to Christian ethics. This approach to work is essential if Christians are to arrive at ethical guidelines for their conduct as workers. But if we want to be faithful to biblical revelation and relevant to the contemporary world of work, we will need to modify this approach at some points and place it in a broader theological framework.

As testified to by the new proposals by various theologians over the past several decades, modifications are needed in the traditional approach both to the influence of a Christian's new life on her work and the consequences of work on the Christian character. Work in the military industry is an example of the first matter. In contrast to the time of the church fathers, our context of the growing militarization of economies,[24] which runs parallel to the progressive impoverishment of the poor, makes it necessary for us today to reflect carefully on work in the military industry. Christians at the turn of the twenty-first century need to consider not only whether they can serve as soldiers (as the church fathers also discussed), but, further, to what extent working (directly or indirectly) for the military industry is compatible with their faith. To be sure, the opinions of Christians on this issue will differ (depending on their valuation of the just-war theory). But they cannot avoid asking, for instance, what consequences should be drawn from the fact that the results of work in the armament industry "have often proved to be one of the most potent causes of war," as Barth wrote not long after World War II.[25]

With respect to the influence of work on Christian character, we would, for instance, have to supplement the negative (and potentially anthropologically misleading) notion that work serves *ad corpus domandum* with the positive idea that human beings achieve fulfillment through work.[26] Ethical reflection on work traditionally done in the framework of the doctrine of sanctification also needs to be supplemented with reflection from the perspective of anthropology. It is imperative, for example, to discuss which forms of work are incompatible with the dignity of human beings as God's free and responsible creatures, and which forms of work develop their personality and which stifle it.

Work and God's Purpose with Creation

Necessary as such modifications in the traditional ethical approach to the problem of work are, they are only one step toward a responsible contemporary theological understanding of work. It is insufficient merely to interpret the biblical statements on work, distill from them transculturally binding ethical principles, and combine them into a consistent statement on how a Christian should work. It is also insufficient to ask what individual Christian doctrines (such as the doctrine of creation and anthropology) imply for our ethical assessment of human work. It is rather necessary to *develop a comprehensive theology of work.*

A theology of work is a dogmatic reflection on the nature and consequences of human work. It does not make ethical theological reflection on human work superfluous but provides it with an indispensable theological framework. For it situates the questions of how one should or should not work, and what one should produce, in the larger context of reflection on the meaning of work in the history of God with the world and on the place of work in human beings' relation to their own nature, to their fellow human beings, and to the natural world.

Why should we not be satisfied with ethical theological reflection on human work? First, the biblical witnesses themselves not only prescribe how human beings should or should not work, but also cast light on the ultimate meaning of human work. They do not consider human work only under the rubric of sanctification, but place it in the broader perspective of God's purposes with creation (see Gen. 1 and 2).[27] Second, the nature and the consequences of human work themselves require a broader horizon of theological reflection on work than the doctrine of sanctification provides. As human beings work, they change themselves as well as their social and natural environment in the course of history. Ethical questions about work can thus be properly addressed only in the context of a broad reflection on the anthropological, social, and cosmological dimensions of work: hence the need to interpret and evaluate work and its consequences from a dogmatic perspective.

Since various and sundry "theologies of . . ." have been mushrooming over the two past decades, two explanations about the character of a theology of work are needed. First, one should not take the theology of work for another "fad theology" (as was, for instance, the theology of the death of God). Since human work is not a fad, theological reflection on it cannot be a fad either. Cultural fads come and go, but work remains as long as human beings remain. In spite of all the changes in the nature of work throughout history, work has been and will continue to be a fundamental condition and dimension of human existence. No theology that wants to take human existence seriously will be able to circumvent theological reflection on human work.

Second, it is important to distinguish carefully between the ways the word "theology" is used in the syntagmas "theology of work" and, for example, "theology of liberation." Theology of liberation elevates liberation to the status of the methodological principle for the whole of theological reflection. Theology of liberation is not a theological reflection on a particular aspect of human life (a genitive theology), but a new way of doing theology as a whole.[28] The task of a theology of work is much more modest. It *is* a genitive theology, for it does not seek to make work the governing theological theme, but to treat it from a dogmatic perspective. The effort to organize theological reflection around the theme of work would be misplaced because it would amount to theological acquiescence to the near-total dominance of work in many contemporary societies. In one crucial respect, such a theology of work would be a mere reflection of the present world of work and thus would forfeit its function as a critical partner in the contemporary discussion about work.

My intention in writing this book is not to add another volume to the flood of ethical theological literature on human work published in recent years (especially on the ethical aspects of the unemployment problem). The purpose of this book is to *develop a new—pneumatological—theology* of work. Especially from Protestant pens, theologies of work are in short supply. Protestant publications on work as a rule ignore the dogmatic perspective on the question of work or assume that nothing more needs to be said about it.[29] This book is written from the persuasion that dogmatic reflection on human work is essential to an adequate theological treatment of the subject, and that there is a host of unresolved questions demanding further investigation; and, above all, that the dominant paradigm for understanding work in Protestant theology is inadequate.

My main task will be to develop a theological framework for understanding human work and to elucidate the implicit ethical principles that should guide our efforts to assess and restructure the world of work. For lack of space and because of my interest and specialization, I have chosen to refrain from making proposals about how these ethical principles should be translated into concrete policies. The complex task of such a translation I take to be the creative assignment of Christian economists and social scientists (to be carried out in dialogue with theologians).

I am aware of the problems involved in keeping safely hidden behind normative theological generalities—not the least of them being the fact that "we have not yet fully understood the claims of any moral philosophy until we have spelled out what its social embodiment would be."[30] But one of the worst ways to avoid these problems would be to rush, theological student that I am, into the economists' and social scientists' fields of competence with policy-shaping proposals that claim to bear the stamp of divine approval. Short of engaging in a truly interdisciplinary project, one of the best ways to avoid this problem is for a theologian to stick to her own discipline and formulate normative principles while taking carefully into account the concrete realities in which these principles have to be implemented. This I have striven to do.

ON CRAFTING A THEOLOGY OF WORK

How does one arrive at a theology of work? In the past, theologians have frequently attempted to formulate a Christian understanding of work by analyzing and combining individual passages of the Bible that speak about human beings and their daily work. The procedure was intended to be strictly inductive, and it resulted in books and articles about *biblical teaching* on work.[31] If one attempts to develop a theology of work in such a way, one runs up against three major problems.

First, the New Testament, the key source for developing a Christian theology of work, addresses the topic of human work only occasionally, and as a subordinate theme at that.[32] The few relevant New Testament passages consist of specific instructions about how Christians should work but make no fundamental affirmations about the meaning of human work. Taken together, these passages simply do not add up to a *theology* of human work. Some Old Testament passages (like Gen. 1 and 2) look more promising at first sight since they include a more comprehensive perspective on work. But they provide us at best only with some elements of a theology of work. Moreover, even these elements are not useful for a *Christian* theology of work just as they stand. To integrate them into a Christian theology, we have to interpret the Old Testament statements on human work in the light of the revelation of God in Christ.[33]

Second, a deep divide separates the world of work in biblical times from work in present industrial and information societies.[34] This ever-widening gap precludes developing a theology of work relevant to our time through the "concordance method" without placing biblical references within a larger theological framework. The explicit biblical statements about work are, for instance, more or less irrelevant to fundamental contemporary questions such as the connection between work or unemployment and human identity, the character of humane work in an information society, and the relationship between work and nature in an age of permanent technological revolution. As Moltmann correctly (given a degree of exaggeration) observes, "anyone who inquires about the work ethos of the Bible runs up against the cultural history of past societies if he or she only investigates the statements on human work."[35] So even if there were enough material in the biblical records to construct a theology of work by the concordance method, a theology of work crafted in such a way would be of limited relevance to the modern world of work.

Third, even when biblical statements about work are applicable to the present, it is still not immediately obvious what significance should be ascribed to each statement in relation to the others and hence, also, precisely how they should inform Christian thinking and behavior. This information is provided by the theological framework in which we place these statements. Representatives of divergent views on economic issues will, for instance, rarely disagree on whether or not the prophet Amos denounced exploitation and the resulting poverty. But they will differ radically about the urgency of combatting exploitation and about the most efficient ways of doing so. The reasons for these differences are manifold, but there is no doubt that one reason is the different overarching theologies in whose light disagreeing Christians read Amos' denunciations.[36] The theological framework is, therefore, crucial for determining the import of individual biblical statements on work.

The inductive approach to developing a theology of work is inadequate because of the scarcity of biblical materials, their limited relevance to the modern world of work, and their ambiguous nature. It is illusory to think that we can treat the biblical statements about work as pieces of a large jigsaw puzzle that we only have to arrange according to the pattern implicit in the pieces themselves in order to get a theology of work that is both biblical and relevant. We need instead to proceed deductively: we need to set up a theological framework in which we then can integrate the biblical statements on work. As a matter of fact, a theological framework is always operative in the interpretation of biblical texts, whether or not the interpreter is conscious of such an influence. Interpretation takes place from within particular theological traditions, which "almost always take the form of *ways of understanding the message of the Bible as a whole*—they take the form of overarching *theologies.*"[37] To develop a theology of work means to consciously place biblical statements about work in the context of a reading of the Bible as a whole and to apply both these individual statements and the overarching reading of the Bible to the contemporary world of work.

I am, of course, not suggesting that one can set up a theological framework independently of the relevant biblical statements. It is an essential characteristic of all authentic Christian "theological puzzles" that the individual pieces that come from the biblical materials also contribute normatively to the shape of the framework of the puzzle. If we cannot arrive at a theology of work simply by combining the individual biblical statements and ethical principles about work that can be derived from them, still less can we do so by manipulating them to fit our preconceived pattern, and certainly not by ignoring them altogether. No aspect of a theology of work is acceptable if it can be biblically patently falsified (granted the difficulties of 'patent falsification' pointed out by philosophers of science).

THEOLOGY OF WORK AND NEW CREATION

The broad theological framework within which I propose to develop a theology of work is the concept of the *new creation*. I will not attempt at this point to give detailed justification for this step or unfold its implications for understanding human work. It should suffice here to say that I am following the basic insight of Moltmann's *Theology of Hope* that at its very core, Christian faith is eschatalogical. Christian life is life in the Spirit of the new creation or it is not Christian life at all. And the Spirit of God should determine the whole life, spiritual as well as secular, of a Christian. Christian work must, therefore, be done under the inspiration of the Spirit and in the light of the coming new creation.[38] The rest of the book consists of the elaboration on this theme. I believe that the reasons for opting for an "eschatalogical" and pneumatological theology of work are best given in the process of its development.

In the remainder of this chapter I want to point out some formal features of a theology of work based on the concept of new creation and indicate how the contemporary reality of work requires a framework of such breadth.

A Christian Theology of Work

The first and most basic feature of a theology of work based on the concept of new creation is that it is a *Christian* theology of work. It is developed on the basis of a specifically Christian soteriology and eschatology, essential to which is the anticipatory experience of God's new creation and a hope of its future consummation.

Under the assumption that Christian faith makes sense, Christian theological reflection about work has an important advantage over secular philosophies of work in that it is furnished with an adequate basis for moral discourse about work. Dostoyevsky (and before him, Nietzsche) rightly maintained that everything is allowed if there is no God. It seems that judgments about right and wrong can be adequately justified only in the context of religious discourse (which does not mean that those who do not accept a religious world view cannot demonstrate exemplary moral qualities, in spite of their lack of adequate justification for their behavior).[39] The belief that reason can establish values is likely to be one of "the stupidest and most pernicious illusion[s]."[40]

But there is also a disadvantage in having a Christian foundation for ethical reflection on work. The problems of work are common problems for all the peoples of the world today. Ours is a pluralistic world, and only a minority of its inhabitants give intellectual assent to Christian beliefs. Even fewer people feel committed to the moral implications of Christian beliefs. Some of my readers who do not share my Christian presuppositions might think that no fruitful exchange of ideas can take place between them and me. So it might seem that I can have the advantage of a solid foundation for ethical reflection only at the price of forfeiting its relevance and persuasiveness to a non-Christian audience.

This is not the place to try to convince non-Christians that Christian faith makes sense (which I believe it does). I only want to remind them of what is commonplace in the philosophy of science; namely, that valid insights can be gained from erroneous and even absurd metaphysical beliefs (which they might believe mine are).

A Christian theologian who takes the concept of new creation as a framework for theological and ethical reflection should, at any rate, be hesitant in qualifying non-Christian moral persuasions as erroneous or absurd *tout court*. This framework does not require a black-and-white view of the world. For the Spirit of God is at work not only in the present anticipation of the new creation in the Christian community, but also in the world. Moreover, in the Christian community the new creation is presently being realized only in an anticipatory form. Since the Last Day is yet to come, a theologian can never pronounce theological judgments from the seat of the Final Judge, so to speak. A theologian's views are not absolute. Furthermore, as she relativizes her own statements, a theologian must always be ready to hear the voice of the Spirit of God in the moral discourse of non-Christians (without forgetting, however, to apply a "hermeneutic of suspicion" here, too). Christian moral discourse is exclusive in the sense that it is based on the concept of new creation ushered in by Christ, but it is also inclusive in the sense that it respects other traditions and is ready to learn from them because it is ready to hear from them also the voice of the Spirit of Christ.[41]

A Normative Theology of Work

A theology of work based on the concept of new creation purports to be a *normative* understanding of work. As I see it, in writing this book, I am not merely stating what I, or anybody else, subjectively considers to be a preferred state of affairs. The book is not primarily about what I desire human work to be. It is also not about how a particular subculture desires to structure its world of work. It is about what human beings *should* desire their work to be.

I believe the principle to be wrong that the "sole evidence it is possible to produce that anything is desirable is that people actually desire it."[42] The principle is incorrect whether one understands it in terms of utilitarianism (as did Mill, who formulated it) or in terms of ethical egoism. Because for Christian theology each individual does not determine his own moral universe, the factuality of desires—either individual desires or common desires reached by consensus—by no means establishes their objective desirability. What people desire is objectively desirable only when it corresponds to what the loving and just God desires for them as God's creatures. And God desires the *new creation* for them. New creation is the end of all God's purposes with the universe, and as such, either explicitly or implicitly is the necessary criterion of all human action that can be considered good. For this reason, normative principles are implied in the concept of new creation, which should guide Christians in structuring the reality of human work.

I am not saying that the desires of the people are *politically* irrelevant but that they are not *ethically* decisive. No one may impose the goals implicit in the concept of new creation either on the majority or a minority of population against their will. Such an imposition would violate the freedom of people and hence be in contradiction with these goals themselves. Human freedom must be respected as an end in itself because it is an essential dimension of human personhood, which is an end in itself. Norms may be politically implemented only when they become public preferences through truly democratic processes (which are not to be confused with a pseudo-democratic civil war of interests in which the majority—whether moral or immoral—wins by use of "civilized" brute force, but must be grounded in public preferences that are based on persuasive public moral discourse). In any case, disregard for the preferences of people in social interaction characterizes a dictatorship, and in that sense it is politically and ethically unacceptable. But the preference satisfaction and its distribution across a community is not what fundamentally matters in the formulation of *ethical* norms.[43]

If anyone is offended by my objectivist approach to ethical discourse, I suggest that she interpret what I consider to be objective normative statements as my own subjective preferences or as the preferences of a subculture to which I belong. From my standpoint this would not so much be a wrong interpretation of my views as an inadequate one. For I am not only talking about what I believe to be good objectively but also about what I and (possibly) some of my fellow Christians subjectively consider to be good. Hence I can accept (with some dissatisfaction) a critical reading of this book that is content to measure the extent to which my subjective preferences here expressed make sense in the framework of the individual or communal moral sensitivity of my reader.

It is crucial to determine more precisely the normative function of the concept of new creation. To this end I would like to draw attention to the traditional ethical distinction between justice and love. The concept of new creation implies certain principles that cannot be set aside if justice is to prevail. This we might call the "ethical minimum." But the new creation also implies principles that point beyond the way of justice to the way of love, which we might call the "ethical maximum." All responsible Christian behavior has to satisfy the ethical minimum and, inspired by the sacrificial love of Christ demonstrated on the cross and guided by the vision of the new creation, move toward the ethical maximum. The ethical minimum is the *criterion* for structuring the world of work, the ethical maximum the necessary *regulative ideal*.

The ethical maximum may not be zealously transmuted from regulative ideal to sacrosanct criterion. As one uses the ethical maximum to optimize structures, one must take soberly into account what is practically realizable.[44] Otherwise one is likely to distort what is meant to be a beneficial critical instance into a tyrannical ideology. At the same time it is crucial not to set love aside as useless in social ethics. Even if one does not operate only with a procedural understanding of justice (as I do not),[45] the practice of justice alone will not be sufficient to create a humane society. For without love, there is no *shalom*.

A Transformative Theology of Work

Since a theology of work has normative ethical implications, its task is not merely to interpret the world of work in a particular way, but to *lead* the present world of work "towards the promised and the hoped-for transformation"[46] in the new creation. To be sure, theological interpretations of work are not pointless; even less should they be simply denounced as a devious attempt to "befog the brain with supernatural, transcendent doctrines."[47] But a theological interpretation of work is valid only if it facilitates transformation of work toward ever-greater correspondence with the coming new creation.

The transformative function of a theology of work demands that, in developing it, we not only attentively read its sources (biblical revelation on work) and carefully analyze the nature of the object of study (the contemporary situation of work), but also reflect critically on the praxis that can follow from the formulations produced by a theology of work. Since it is impossible to make theological statements that cannot be misused, however, it may seem unjust to require a theology of work to take into account the potential consequences of its theological formulations; that requirement seems to put a theology at the mercy of people's willingness or unwillingness to understand and practice it properly. But the point is surely not to pay attention to the interpretive whims of individuals, but to take into consideration broad tendencies toward misinterpretation that are rooted in the logic of the cultural context in which theological statements are uttered.

It might also be objected that concern for the practical consequences of theological formulations easily degenerates into an approach that takes these consequences as an independent basis for theologizing. The purpose of critical reflection on the function of theology, however, is not to make the desired reflection of theological formulations determine their content, but to ensure that theological formulations serve the function that their content dictates.

In facilitating a transformation of human work, a theology of work cannot operate with an *evolutionist* understanding of social realities. The concept of new creation precludes all naïve belief in the permanence of human moral progress. A truly *new* creation can never result from the action of intrahistorical forces pushing history toward ever-superior states. Although we must affirm the continuity between present and future orders,[48] that affirmation should not deceive us into thinking that God's new creation will come about in linear development from the present order of things. The implied normativeness of new creation enables us to evaluate (and appreciate) present achievements of the human race, and the radical newness of God's future creation frees us from having to press history into a utopian developmental scheme. Holding to the theological framework of the new creation allows us to perceive progress in certain aspects of social life or in certain historical periods, and it allows us also at other times to share Luther's view that "the world [as a whole, or a particular 'world'] is deteriorating from day to day."[49] The concept of the new creation allows us to *combine* the normative approach to social life with what might be called a "kaleidoscope" theory of social life, according to which social arrangements shift in various ways under various influences (divine, human, or demonic) without necessarily following an evolutionist or involutionist pattern.

A Comprehensive Theology of Work

A theology of work based on the concept of new creation needs to be *comprehensive*. Since the new creation is a universal reality (the creation of a new heaven and a new earth), a theology of work based on it needs to answer the question of how human work is related to all reality: to God, human beings, and their nonhuman environment.

Such an all-inclusive framework for a theology of work is demanded by the significance of human work itself. It is possible to see the whole of human history as a result of the combined work of many generations of human beings. Human work, properly understood theologically, is related to the goal of all history, which will bring God, human beings, and the nonhuman creation into "shalomic" harmony. Neither personal development (self-realization) nor communal well-being (solidarity) alone are adequate contexts for a theological reflection on human work. To do justice to the nature of its subject (work) and its source (Christian revelation), a theology of work must investigate the relation of work to the future destiny of the whole creation, including human beings as individual and social beings, and the nonhuman environment. The appropriate theological framework for developing a theology of work is not anthropology, but an all-encompassing eschatology.

Because of the universality of new creation, a theology of work needs to be comprehensive by relating work, not only to all dimensions of reality—God, human beings, and nature—but also to humanity and nature in their entirety. It needs to be a *global* theology.[50] Because the world of human work is a global world, a theology of work must attempt to reflect on work in a global context.

The interdependence created by the first division of labor at the dawn of history has grown to include almost the entire human race and shows a tendency toward further increase. The wealthy and technologically developed North and the destitute South are increasingly dependent on one another for resources; the North being unable to function without the South's raw materials and markets, the South needing the North for technology and know-how. The emerging world economy is transforming our world from a set of self-sustaining tribes and nations into a global village (or a global city). The unity of the human race is no longer merely an abstract notion. The same is true of the natural environment: powerful technology (created both *for* the emerging world economy and *by* it) has had the ecological effect of fusing more or less self-contained geographical units into a single global environment.

A theology of work must be comprehensive not only in the synchronic sense (given the global village and environment) but also in a diachronic sense. As a network of interdependently working individuals and communities, the present-day generation is unalterably shaping the world for future generations. Through the cumulative effects of modern technology on the human environment, a new world is being created, a new world that is potentially no less a nightmare-world than a dream-world. For this reason we need a theology of work that reflects on the present situation in view of the future that the present is giving birth to. If we base a theology of work on the concept of new creation, we can think of the work of the present generation and that of coming generations as two aspects of one reality, as the combined work of the single human race. A theology of work adequate to the modern world of work must be cross-cultural and cross-historical, a pan-human theology of work.[51]

A Theology of Work for Industrial Societies

A theology of work based on the concept of new creation is open to the contributions of *individual cultural units*. New creation is a universal reality that realizes itself in history through the Spirit of God. It does not destroy history or obliterate the diversity of the individual cultures it includes. The new creation is mediated in different ways in different cultures. It is well known, for instance, that what people think and feel about work, the extent to which work is gratifying, frustrating, or merely endurable to them, depends at least partly on the particular culture in which they live.[52] This variety of cultural forms and their partial preservation in the new creation implies that a diversity of valid theologies of work conditioned partly by the character and the understanding of work in a given culture could exist.

To acknowledge diverse theologies of work is not to succumb naïvely to cultural and historical relativism. Such relativism is not only philosophically problematic;[53] in Christian theology it would be out of place. For the notion of new creation implies universally valid normative principles. Some aspects of work that seem meaningful in a particular culture will be at odds with these principles. One can imagine that the work of some slaves was gratifying to them (the "pleasant slavery" Marx spoke of[54]), yet as a mode of economic arrangement of interhuman relations, slavery is clearly morally unacceptable. When culture conflicts with new creation, it is culture that has to go. There are, however, aspects of human work that are ethically neutral, but valuable nevertheless as expressions of a particular culture (such as some technological and aesthetic aspects of work). Such culturally conditioned ways of doing work could be validly integrated into the concept of meaningful work. The normative character of a theology of work does not preclude different *accents* specific to a particular culture in a theology of work constructed for that culture.

My concern, however, is not so much with the culturally specific aspects of a theology of work as with its normative claims and with their realization. Reflection on the realization of the normative principles must take into consideration the specific situation in which those principles are to be realized. In that sense, too, a responsible theology of work will necessarily be colored by the character and the understanding of work in the contemporary societies for which it is being developed.

This book deals with the reality and understanding of work of industrializing and industrialized societies, which are experiencing a slow but irreversible transformation into information societies. These include not only the societies of the so-called First World but also some of the developing societies of the Second and Third Worlds (for example, my home country, Yugoslavia). The conditions and the character of work in industrial societies are becoming increasingly characteristic of work elsewhere as a genuine world economy is being created.

I am writing, therefore, for a particular context (industrializing and industrialized societies), which nevertheless has a universalizing tendency. Since a theology of work should have a global character, I will also try to bear in mind the implications of my proposals for presently existing societies or segments of societies still in the pre-industrial phase. Often a tension is felt between a particular context and the universal outlook. This tension can never be removed completely, but we can and should strive to reduce it. A dialogue between thinkers from different contexts who nevertheless have a universal perspective could facilitate clearer formulation of the tensions between all the different particular interests and thus contribute toward their reduction. This in turn is a precondition for peaceful living in a world whose inhabitants are growing increasingly interdependent.

CHAPTER 4

Work, Spirit, and New Creation

The foregoing analysis of the dominant theories of work in the world today and of the changing present reality of work (Chapters 1 and 2) served to place the object of our study in focus. I then pleaded in Chapter 3 for developing a theology of work instead of an ethic of work and indicated some formal features of a proposed pneumatological theology of work cast within the framework of the concept of "new creation."

In the present chapter, I will lay a foundation for such a theology of work and sketch its basic contours. The first major section deals with the ultimate *significance* of work by discussing the question of the continuity or discontinuity between the present and the eschatological orders, and with the fundamental *meaning* of work by arguing in favor of understanding work as cooperation with God. In the second major section I will first give reasons why a pneumatological theory of work is possible. Then, in a critical dialogue with the dominant Protestant view of work as vocation developed within a protological framework, I will argue for a pneumatological understanding of work based on a theology of *charisms*, which suggests that the various activities human beings do in order to satisfy their own needs and the needs of their fellow creatures should be viewed from the perspective of the operation of God's Spirit. I will end by defending the pneumatological understanding of work from a possible criticism of being a Christian ideology of work.

WORK AND NEW CREATION

The question of continuity or discontinuity between the present and future orders[1] is a key issue in developing a theology of work. The ultimate significance of human work depends on the answer to this question, for it determines whether work as occupation with transitory things and relations (*vita activa*) has an inherent value or whether it merely has instrumental value as a means to make possible the occupation with eternal realities (*vita contemplativa*).[2]

Eschatology and the Significance of Human Work

If we leave aside the more modern—and in my view theologically and religiously not very persuasive—ethical and existential interpretations of the cosmological eschatological statements, Christian theologians have held two basic positions on the eschatological future of the world. Some stressed radical discontinuity between the present and the future orders, believing in the complete destruction of the present world at the end of the ages and creation of a fully new world. Others postulated continuity between the two, believing that the present world will be transformed into the new heaven and new earth. Two radically different theologies of work follow from these two basic eschatological models.

Work and the Annihilatio Mundi

If the world will be annihilated and a new one created *ex nihilo*, then mundane work has only earthly significance for the well-being of the worker, the worker's community, and posterity—until the day when "the heavens will pass away with a loud noise, and the elements, will be dissolved with fire" (2 Pet. 3:10). Since the results of the cumulative work of humankind throughout history will become naught in the final apocalyptic catastrophe, human work is devoid of direct ultimate significance.

Under the presupposition of eschatological *annihilatio mundi*, human work can, of course, indirectly serve certain goals whose importance transcends the death of either the individual or the whole cosmos. One can, for instance, view work as a school for the purification of the soul in preparation for heavenly bliss. Christian tradition always insisted on the importance of work for individual sanctification.[3] For, as Thomas Aquinas put it, it removes "idleness whence arise many evils" and "curbs concupiscence."[4] One can also maintain (as Karl Barth did) that work is indirectly ultimately significant because it keeps the body and the soul together, thus enabling Christian faith and service: in order to believe and serve, human beings have to live, and in order to live, they have to work.[5] According to such views, human work and its results are necessary, for without them the Christian *opus proprium* (faith, sanctification, or service) cannot take place. Yet, being merely prerequisites for this *opus proprium*, human work and its results are eschatologically insignificant independent of their direct or indirect influence on the souls of men and women.

When one refuses to assign eschatological significance to human work and makes it fully subservient to the vertical relation to God, one devalues human work and Christian cultural involvement (I use the word in the broad sense inclusive of social and ecological involvement). It is, of course, logically compatible both to affirm that the world will be annihilated at the end and at the same time to strive to improve the life of individuals, to create adequate social structures, and even to be motivated to care effectively for the environment. There is nothing contradictory in wanting to use the world and delight in it as long as it lasts (or as long as human beings last in it). Because it is possible to affirm enjoyment in the world while believing in its destruction, it is also possible to consider one's cultural involvement as a way of integrally loving one's neighbor. If Bach, for instance, were annihilationist, should he have had qualms about composing his music?[6] Of course not. He could have done this out of a desire to spiritually elevate his audience and thereby glorify God.[7]

Belief in the eschatological annihilation and responsible social involvement are logically compatible. But they are *theologically inconsistent*. The expectation of the eschatological destruction of the world is not consonant with the belief in the goodness of creation: what God will annihilate must be either so bad that it is not possible to be redeemed or so insignificant that it is not worth being redeemed. It is hard to believe in the intrinsic value and goodness of something that God will completely annihilate.

And *without a theologically grounded belief in the intrinsic value and goodness of creation, positive cultural involvement hangs theologically in the air.* Hence Christians who await the destruction of the world (and conveniently refuse to live a schizophrenic life) shy away as a rule—out of theological, not logical, consistency—from social and cultural involvement. Under the presupposition that the world is not intrinsically good, the only theologically plausible justification for cultural involvement would be that such involvement diminishes the suffering of the body and contributes to the good of the soul (either by making evangelism possible or by fostering sanctification). Comfort, skill, or beauty—whether it is the beauty of the human body or of some other object—could have no more intrinsic value than does the body itself; they could be merely a means to some spiritual end. To return to our example, even if annihilationist presuppositions need not discourage Bach's work, his composing in order for people to take pleasure in his music could not be theologically motivated. He would have no theological reason for this important way of loving others. This problem would not arise, however, if Bach believed in the intrinsic goodness of creation. And he could do this only if he believed in the eschatological transformation rather than destruction.

Work and the Transformatio Mundi

The picture changes radically with the assumption that the world will end not in apocalyptic destruction but in eschatological transformation. Then the results of the cumulative work of human beings have intrinsic value and gain ultimate significance, for they are related to the eschatological new creation, not only indirectly through the faith and service they enable or sanctification they further, but also directly: the noble products of human ingenuity, "whatever is beautiful, true and good in human cultures,"[8] will be cleansed from impurity, perfected, and transfigured to become a part of God's new creation. They will form the "building materials" from which (after they are transfigured) "the glorified world" will be made.[9]

The assurance of the continuity between the present age and the age to come (notwithstanding the abolition of all sinfulness and transitoriness that characterize the present age) is a "strong incentive to . . . cultural involvement."[10] For the continuity guarantees that no noble efforts will be wasted. Certainly, cultural involvement is not the most important task of a Christian. It would indeed be useless for a woman to conquer and transform the world through work but through lack of faith lose her soul (see Mark 8:36). Yet, as faith does not exist for the sake of work (though it should stimulate, direct, and limit work), so also work does not exist merely for the sake of faith (though one of its purposes is to make faith possible). Each in its own way, faith and human work should stand in the service of the new creation. Not that the results of human work should or could create and replace "heaven." They can never do that; though, charmed with success, people often forget that simple truth. Rather, after being purified in the eschatological *transformatio mundi*, they will be integrated by an act of divine transformation into the new heaven and the new earth. Hence the expectation of the eschatological transformation invests human work with ultimate significance. Through it human beings contribute in their modest and broken way to God's new creation.

The ascription of intrinsic value and ultimate significance to positive cultural involvement is not the only benefit of developing a theology of work within the framework of belief in eschatological continuity. In addition, such a belief gives human beings important inspiration for action when their efforts at doing good deeds, at finding truth about some aspect of reality, and at creating beauty are not appreciated. The question is not merely whether Bach would have qualms about composing music if he were an annihilationist. The question is also whether all those unappreciated small and great Van Goghs in various fields of human activity would not draw inspiration and strength from the belief that their noble efforts are not lost, that everything good, true, and beautiful they create is valued by God and will be appreciated by human beings in the new creation.

The New Testament on the Significance of Work

It might seem that discussing eschatological annihilation and transformation is a roundabout way to reflect theologically on the significance of human work. Should not the explicit New Testament statements about work determine our perspective on the issue? If they did, we would come to a rather different valuation of cultural involvement than the one implied in the idea of *transformatio mundi*. For we search in vain in the New Testament for a cultural mandate, let alone for the "gospel of work."[11] Jesus left carpenter's tools when he started public ministry, and he called his disciples away from their occupations. Only indirectly did he affirm the need to work: when he said that people will be judged on the basis of their efforts to satisfy basic human needs of the poor (food, drink, clothing; Matt. 25:34ff.).[12] Later, we find in the epistles an explicit command to work, but with the clear specification that work should serve the needs of the workers and their neighbors (see 2 Thess. 3:6ff.; Eph. 4:28). The explicit New Testament statements about work view it very soberly as a means of securing sustenance, not as an instrument of cultural advancement.

The key question is how to interpret the silence of the New Testament about the possible broader significance of human work. Is it an implicit discouragement of cultural involvement or merely an expression of a singleminded *concentration* on a different kind of work needed in a particular period of salvation history (see Matt. 9:37f.)? In answering this question it is good to remember that in the Old Testament, the "scripture" for the early Christians, the purpose of work was not merely sustenance, but also cultural development, which included activities ranging from perfecting building techniques to the refinement of musical skills (See Gen. 4:17ff.). Moreover, Genesis views the diversification of employments required by such cultural development as a result of divine blessing.[13] The Old Testament view of work should caution us against concluding too hastily that a positive valuation of cultural development is incompatible with a New Testament understanding of Christian faith.

Important as this argument from the Old Testament is, it is not decisive. The answer to the question of how to translate into a Christian theology of work the silence of the New Testament about any broader significance of work than mere sustenance depends ultimately on the nature of New Testament eschatology. For the significance of secular work depends on the value of creation, and the value of creation depends on its final destiny. If its destiny is eschatological transformation, then, in spite of the lack of explicit exegetical support, we *must* ascribe to human work inherent value, independent of its relation to the proclamation of the gospel (human work and the proclamation of the gospel are each in its own way directed toward the new creation). Since much of the present order is the result of human work, if the present order will be transformed, then human work necessarily has ultimate significance. The interpretation of the explicit New Testament statements about the significance of work depends, therefore, on the eschatological framework in which they are set. So the search for a direct answer in the New Testament to the question about a possible broader significance of work than securing sustenance leads us to return to our initial discussion about the continuity between the present and future orders.

Eschatological *Transformatio Mundi*

Both explicit and implicit theological arguments can be adduced for the idea of the eschatological *transformatio mundi* and hence for the continuity between the present and the future orders.

Kingdom of God for This Earth

One can argue indirectly for the eschatological transformation of the world instead of annihilation by pointing to the *earthly locale of the kingdom of God*.[14] R. H. Gundry has argued persuasively that in Relevation the saints' dwelling place is the new earth. It is "quite clear that the Book of Revelation promises eternal life on the new earth . . . , not ethereal life in the new heaven." In correspondence to the saints' earthly dwelling place, the promise to the church at Smyrna—"but you are rich" (2:9)—calls for a "materialistic reading": it refers to "a redistribution of property . . . to the saints." Moreover, Relevation complements the economic aspect of the promise by adding a political aspect: the saints will rule as "new kings of the earth, all of them, the whole nation of kings."[15]

The same emphasis on the new earth as the eschatological dwelling place of God's people found in Revelation is also present in the Matthew gospel. The prayer for the coming of the kingdom (6:10) is a prayer for God's "rule over all the earth," and seeking the kingdom (6:33) "means desiring the final coming of his rule on earth."[16] Similarly, the "earth" in the promise of inheriting the earth given to the meek (5:5) can only refer to "the earthly locale of God's kingdom."[17] In the *eschaton*, the resurrected people of God will inhabit the renewed earth.

The stress on the earthly locale of the kingdom of God in the New Testament corresponds not only to the earthly hopes of the Old Testament prophets (Isa. 11:6-10; 65:17-25), but even more significantly to the Christian doctrine of the resurrection of the body. Theologically it makes little sense to postulate a nonearthly eschatological existence while believing in the resurrection of the body.[18] If we do not want to reduce the doctrine of the resurrection of the body to an accidental part of Christian eschatology, we will have to insist (against Thomas Aquinas, for instance) that perfect happiness does depend on the resurrected body.[19] And if the concept of "body" is not to become unintelligible by being indistinguishable from the concept of the "pure spirit," we must also insist that "external goods" are necessary for perfect happiness.[20] The resurrection body demands a corresponding glorified but nevertheless material environment. The future *material* existence therefore belongs inalienably to the Christian eschatological expectation.[21]

Liberation of Creation

Some New Testament statements explicitly support the idea of an eschatological *transformatio mundi* and indicate that the apocalyptic language of the destruction of "all these things" (2 Pet. 3:11) should not be taken to imply the destruction of creation. In Romans 8:21 Paul writes that the "creation itself . . . will be set free from its bondage to decay and obtain the glorious liberty of the children of God." The liberation of creation (i.e. of "the sum-total of sub-human nature, both animate and inanimate"[22]) *cannot occur through its destruction but only through its transformation.* As F. F. Bruce rightly points out, "if words mean anything, these words of Paul denote not the annihilation of the present material universe on the day of revelation, to be replaced by a universe completely new, but the transformation of the present universe so that it will fulfill the purpose for which God created it."[23] When God ushers in his final kingdom, the striving of "everything in heaven and on earth . . . after renewal" will be fulfilled.[24]

The biblical statements that affirm continuity between the present and future orders are theologically inseparable from the Judeo-Christian belief in the goodness of divine creation. The belief in the continuity between the present and the new creation is an eschatological expression of the protological belief in the goodness of creation; you cannot have one without having the other. It makes little sense to affirm the goodness of creation and at the same time expect its eschatological destruction. And goodness is a predicate not only of the original but also of the present creation, the reality of evil in it notwithstanding. God cannot, therefore, ultimately "reject" creation, but will—as we read in the Pastorals in relation to food—"consecrate" it (1 Tim. 4:4).

It is, of course, possible to believe that the goodness of the material creation is merely instrumental, in which case eschatological annihilation would not deny the goodness of creation. Like food, all material objects would be good because they are necessary for keeping the human body alive, and the human body would be good because it provides a temporary dwelling place for the soul. Alternatively, one can posit the instrumental goodness of the material creation by affirming that it is only a temporary means of manifesting God's greatness and glory. There is no reason to deny or denigrate the instrumental goodness of the material creation. But the material creation is more than a means; it is also an end in itself. For one, we encounter in the biblical texts what might best be described as the "soteriological independence" of the material creation: creation too will participate in the liberty of the children of God (Rom. 8:21; see Gen. 9:10ff).[25] Furthermore, anthropologically we have to maintain that human beings do not only have a body; they also are body.[26] It follows that the goodness of the whole material creation is intrinsic, not merely instrumental. And the belief in the intrinsic goodness of creation is compatible only with the belief in the eschatological continuity.

Human Works in the Glorified World?

Belief in the eschatological transformation of the world gives human work special significance since it bestows independent value on the results of work as "building materials" of the glorified world. As I have shown above, it makes theological sense to talk about human contribution to the glorified world. But is such talk logically plausible? Is it not a contradiction to ascribe eternal permanence to what corruptible human beings create?[27] A chair becomes broken in a year, bread is eaten in a day, and a speech forgotten in an hour.

Most of the results of human work will waste away before they see the day of eschatological transformation.

We should not think only in terms of the work of isolated individuals, however, but also of the cumulative work of the whole human race. The work of each individual contributes to the "project" in which the human race is involved. As one generation stands on the shoulders of another, so the accomplishments of each generation build upon those of the previous one. What has wasted away or been destroyed often functions as a ladder that, after use, can be pushed aside.

Second, although on the one hand much of human work serves for sustenance and its results disappear almost as soon as they have appeared, on the other hand, much human work leaves a permanent imprint on natural and social environments and creates a home for human beings without which they could not exist as human beings. Even if every single human product throughout history will not be integrated into the world to come, this home as a whole will be integrated.

Third, work and its perceived results define in part the structure of human beings' personality, their identity.[28] Since resurrection will be not a negation but an affirmation of human earthly identity, earthly work will have an influence on resurrected personality. Rondet rightly asks whether Gutenberg in a glorified state would be Gutenberg apart from any eschatological relation to the discovery that made him famous.[29] One might go on to ask whether all human beings who have benefited from Gutenberg's discovery would in their glorified state be the same without his discovery. It could be argued that such an understanding of the direct ultimate significance of human work is also possible if one holds to the doctrine of the annihilation of the world. Strictly speaking, this is true. But it seems inconsistent to hold that human creations are evil or insignificant enough to necessitate their destruction and that their influence on human personality—which should be carefully distinguished from the influence the process of work has on the individual's sanctification—is good enough to require eschatological preservation.

It is plausible that the statement in Revelation about the saints resting "from their labors (*kopōn*), for their deeds (*erga*) follow them" (Rev. 14:13; cf. Eph. 6:8) could be interpreted to imply that earthly work will leave traces on resurrected personalities. Since the preservation of the results of work is not in view in this passage, it seems that their deeds can follow the saints only as part and parcel of their personality.[30] Human work is ultimately significant not only because it contributes to the future environment of human beings, but also because it leaves an indelible imprint on their personalities.

Cooperatio Dei

In the past few centuries Christian theologians have come to view human work as *cooperation with God*. In both Protestant and Roman Catholic traditions there is agreement today that the deepest meaning of human work lies in the cooperation of men and women with God.[31] To view work as cooperation with God is compatible with belief in eschatological annihilation (one cooperates with God in the preservation of the world until its final destruction); belief in the eschatological transformation, however, is not only compatible with such a view of work—it requires it.

Depending on how we conceive of human cooperation with God in work, we can differentiate between two types of theologies of work. The one rests on the doctrine of creation and sees work as cooperation with God in *creatio continua*, the other rests on the doctrine of the last things and sees work as cooperation with God in anticipation of God's eschatological *transformatio mundi*. I will take a brief, critical look at both ways to understand human work as cooperation with God.

Cooperation in Preservation

The first way of interpreting work as cooperation with God starts with the Old Testament, especially the creation accounts. The first chapters of Genesis portray human beings, even in their mundane work, as partners with God in God's creative activity. True, the Old Testament stresses the uniqueness of God's act of original creation. No human work corresponds to divine creation *ex nihilo* (*bara*). At the same time it draws an analogy between divine making (*asa*) and human work,[32] which seems to suggest that there is a partnership between the creating God and working human beings.

The second account of creation portrays this partnership in the most vivid manner. While giving the reason for the lack of vegetation on earth, it addresses the relation of God's creation and human work: "For the Lord God had not caused it to rain upon the earth; and there was no man to till the ground" (Gen. 2:5). The growth of vegetation demands cooperation between God, who gives rain, and human beings, who cultivate the ground. There is a mutual dependence between God and human beings in the task of the preservation of creation. On the one hand, human beings are dependent upon God in their work. As the psalmist says, "Unless the Lord builds the house, those who build it labor in vain" (Ps. 127:1a; cf. Ps. 65:11-13). On the other hand, God the Creator chooses to become "dependent" on the human helping hand and makes human work a means of accomplishing his work in the world. As Luther said, human work is "God's mask behind which he hides himself and rules everything magnificently in the world."[33]

As Luther's statement indicates, cooperation with God need not be a conscious effort on the part of human beings. In other words, work must objectively correspond to the will of God, but it need not be done subjectively as God's will. According to biblical records, God made even those whom he will later judge for the work they have done cooperate with him in accomplishment of his will (see Isa. 37:26ff.). Furthermore, cooperation with God can even occur through alienated forms of work, if the results are in accordance with God's will. Although the concept of new creation does demand striving to overcome alienation in work,[34] such nonalienated work is not a necessary precondition of human cooperation with God.

Cooperation in Transformation

There is another, more recent, theological tradition that bases theology of work on human proleptic cooperation in God's eschatological *transformatio mundi*. It includes the essential elements of the understanding of work as cooperation with God in preservation of creation and places them in the eschatological light of the promised new creation. True, the world is presently under the power of sin and is transitory. For that reason human work cannot create God's new world, no matter how noble human motives might be.[35] The description of the "New Jerusalem"—the new people of God—in Revelation makes this plain.[36] The New Jerusalem is the city (which stands for "the people") of *God* and comes "down out of heaven" (Rev. 21:2; cf. 1 Pet. 1:4; Matt. 25:34). As a divine creation it is "a living hope" freed from all evil and corruptibility, and it infinitely transcends everything human beings can plan or execute.[37] The origin and character of the "New Jerusalem" show that the new creation as a whole is fundamentally a gift, and the primary human action in relation to it is not doing but "waiting" (2 Pet. 3:12; cf. Matt. 6:10; Rev. 22:17).

But one should not confuse waiting with inactivity. In the New Testament the injunction to wait eagerly for the kingdom is not opposed to the exhortation to *work diligently for the kingdom*. "Kingdom-participation" is not contrary, but complementary, to "kingdom-expectation" and is its necessary consequence.[38] Placed in the context of kingdom-participation, mundane human work for worldly betterment becomes a contribution—a limited and imperfect one in need of divine purification—to the eschatological kingdom, which will come through God's action alone. In their daily work human beings are "co-workers in God's kingdom, which completes creation and renews heaven and earth."[39]

It might seem contradictory both to affirm human contribution to the future new creation and to insist that new creation is a result of God's action alone. The compatibility of both affirmations rests in the necessary distinction between God's eschatological action *in* history and his eschatological action *at the end* of history. Through the Spirit, God is already working in history, using human actions to create provisional states of affairs that anticipate the new creation in a real way. These historical anticipations are, however, as far from the consummation of the new creation as earth is from heaven. The consummation is a work of God alone. But since this solitary divine work does not obliterate but transforms the historical anticipations of the new creation human beings have participated in, one can say, without being involved in a contradiction, that human work is an aspect of active anticipation of the exclusively divine *transformatio mundi*.

Both the protological and the eschatological understanding of cooperation with God in daily work briefly analyzed above are valid theologically. One can develop a biblically responsible theology of work by using either an eschatological or a protological framework. For a number of reasons, however, I prefer the eschatological framework. Some of the reasons will become clear later, both through the critique of Luther's notion of vocation (which he developed in a protological framework) and through the reasons to be given for the proposed pneumatological understanding of work.[40] Here I want to mention only four.

First, the eschatological nature of Christian existence makes it impossible, to my mind, to develop a theology of work simply within the framework of the doctrine of creation (protological framework).[41] The second reason is the nature of the relation between the first and the new creation. True, because of the eschatological continuity, the new creation is not simply a negation of the first creation but is also its reaffirmation. For this reason we cannot construe a theology of work apart from the doctrine of creation.[42] The new creation is, however, not a mere restoration of the first creation. "The redemption of the world, and of mankind, does not serve only to put us back in the Garden of Eden where we began. It leads us on to that further destiny to which, even in the Garden of Eden, we were already directed."[43] For this reason the doctrine of creation as such is an insufficient basis for developing a theology of work. It needs to be placed in the broad context of the (partial) realization and of the expectation of the new creation. A proponent of an eschatological theology of work will, therefore, not treat the protological and eschatological understandings of *cooperatio Dei* as alternatives. Rather, they complement each other. Since the new creation comes about through a transformation of the first creation, cooperation with God in the preservation of the world must be an integral part of cooperation with God in the transformation of the world.

The third reason for preferring the eschatological to the protological framework is the conceptual inadequacy of "protological" theologies of work for interpreting the modern work. For them, the ultimate purpose of human work as cooperation with God is the *preservation* of the world. Although much of human work has still the purpose of preserving workers and the world they live in, by using powerful modern technology, human beings not only maintain the world as their home but also radically alter the face of the earth. Modern work transforms the world as much as it preserves it, and it preserves it only by transforming it. The static framework of preservation cannot adequately incorporate this dynamic nature of human work (unless we use the framework of preservation to radically call into question the present results of human work and limit the purpose of work to sustenance). Finally, the protological theologies of work tend to justify the *status quo* and hinder needed change in both microeconomic and macroeconomic structures by appealing to divine preservation of the world: as God the Creator preserves the world he has created, so also human beings in their work should strive to preserve the established order.

WORK AND THE DIVINE SPIRIT
A Pneumatological Theology of Work?

One cannot talk about the new creation without referring to the Spirit of God. For the Spirit, as Paul says, is the "firstfruits" or the "down payment" of the future salvation (see Rom. 8:23; 2 Cor. 1:22) and the present power of eschatological transformation in them. In the Gospels, too, Spirit is the agent through which the future new creation is anticipated in the present (see Matt. 12:28). Without the Spirit there is no experience of the new creation! A theology of work that seeks to understand work as active anticipation of the *transformatio mundi* must, therefore, be a *pneumatological* theology of work.

Work and the Spirit

But what does the Spirit of God have to do with the mundane work of human beings? According to most of Protestant theology, very little. It has been "inclined to restrict the activity of the Spirit to the spiritual, psychological, moral or religious life of the individual."[44] One can account for this restriction by two consequential theological decisions. To use traditional formulations: first, the activity of the Spirit was limited to the sphere of salvation, and second, the *locus* of the present realization of salvation was limited to the human spirit. In another part of this book I will try to show that the Spirit of God is not only *spiritus redemptor* but also *spiritus creator*.[45] Thus when the Spirit comes into the world as Redeemer he does not come to a foreign territory, but "to his own home" (John 1:12)[46]—the world's lying in the power of evil notwithstanding. Here, however, I want to discuss briefly the limitation of the Spirit's salvific operation on the human spirit. For my purposes, this is the crucial issue. The question of whether one can reflect on human work within the framework of the concept of the new creation and develop a pneumatological theology of work depends on the question of whether the Spirit's salvific work is limited to the human spirit or extends to the whole of reality.

The exclusion of the human body and materiality in general from the sphere of salvation in Protestant thought[47] is well illustrated by Luther's *The Freedom of a Christian*, a "small book" that, in Luther's own opinion, nevertheless contained his view of "the whole of Christian life in a brief form."[48] Later Protestant theologians have followed Luther rather closely in regard to the materiality of salvation.[49] In *The Freedom of a Christian* Luther makes the well-known distinction between the "inner man" and the "outward man." For the discussion of the materiality of salvation it is crucial to determine what, exactly, Luther means by these expressions. The matter is not as simple as it looks, because he equivocates and makes a *twofold* distinction in his use of those terms.[50]

First, and most obviously, Luther makes an *anthropological* distinction. The exact nature of this anthropological distinction is not easy to establish. In particular, it is not clear what he means by the "inner man." Fortunately, Luther is very clear on what he means by the "outward man": it is the aspect of the human being that is sick or well, free or imprisoned, that eats or hungers, drinks or thirsts, experiences pleasure or suffers some external misfortune.[51] The outward man is a person with respect to his bodily existence in the world. That leaves the inner man stripped of all corporeality as "the naked self which exists concealed in his [human being's] heart."[52] Whatever "the naked self," or as Luther says, the "soul," is, one thing is certain: for Luther it does not denote a human being's bodily existence.

Superimposed on the anthropological distinction between inner and outward man is the second, *soteriological* distinction between "new man" and "old man." Significant for the study of the materiality of salvation is the fact that Luther applies the soteriological distinction between new and old only to the inner man. *"Outward man" is and* (until the day of the resurrection of the dead) *will remain "old man"*—in the case of both the Christian and the non-Christian. Only the inner man can become a new man. The anthropological *locus* of salvation is the inner man.[53] The outward man and the whole material reality remain outside the sphere of the salvific activity of God.[54]

We need to look no further than the Gospels to see that the exclusion of materiality from the sphere of the present salvific activity of the Spirit is exegetically and theologically unacceptable. The Gospels widely use soteriological terminology (e.g., the term *sōzein*) to designate deliverance from the troubles and dangers of bodily life.[55] More significantly, they portray Jesus' healing miracles as signs of the inbreaking kingdom.[56] As deeds done in the power of the Spirit, healings are not merely symbols of God's future rule, but are anticipatory realizations of God's present rule. They provide tangible testimony to the materiality of salvation; they demonstrate God's desire to bring integrity to the whole human being, including the body, and to the whole of injured reality.[57] In a broken way—for healed people are not delivered from the power of death—healings done here and now through the power of the Spirit illustrate what will happen at the end of the age when God will transform the present world into the promised new creation.

When the ascended Christ gave the Spirit, he "released the power of God into history, power which will not abate until God has made all things new."[58] The Spirit of the new creation cannot be tied to the "inner man." Because the whole creation is the Spirit's sphere of operation, the Spirit is not only the Spirit of religious experience but also the Spirit of worldly engagement. For this reason it is not at all strange to connect the Spirit of God with mundane work. In fact, an adequate understanding of human work will be hardly possible without recourse to pneumatology.[59]

Work and Charisms

In a sense, a pneumatological understanding of work is not new. There are traces of it even in Luther. He discussed the *vocatio externa* not only in the context of the Pauline concept of the Body of Christ (which is closely related to Paul's understanding of charisms) but also—and sometimes explicitly—in the context of the gifts of grace: "Behold, here St. Peter says that the *graces and gifts of God* are not of one but of varied kind. Each one should understand what his gift is, and practice it and so be of use to others."[60]

In recent years authors from various Christian traditions have suggested interpreting human work as an aspect of charismatic life.[61] The document of the Vatican II *Gaudium et spes* contains probably the most notable example of a charismatic interpretation of Christians' service to their fellow human beings through work: "Now, the gifts of the Spirit are diverse. . . . He summons . . . [people] to dedicate themselves to the earthly service of men and to make ready the material of the celestial realm by this ministry of theirs."[62] To my knowledge, however, no one has taken up these suggestions and developed them into a consistent theology of work.

The pneumatological understanding of work I am proposing is an heir to the vocational understanding of work, predominant in the Protestant social ethic of all traditions.[63] Before developing a pneumatological understanding of work, it is therefore helpful to investigate both the strengths and weaknesses of the vocational understanding of work. Similarly to any other theory, a particular theology of work will be persuasive to the extent that one can show its theological and historical superiority over its rivals.

Work as Vocation

Both Luther and Calvin, each in his own way, held the vocational view of work. Since Luther not only originated the idea but also wrote on it much more extensively than Calvin, I will develop my theology of work in critical dialogue with Luther's notion of vocation (which differs in some important respects from Calvin's,[64] and even more from that of the later Calvinists).

The basis of Luther's understanding of vocation is his doctrine of justification by faith, and the occasion for its development, his controversy with medieval monasticism. One of Luther's most culturally influential accomplishments was to overcome the monastic reduction of *vocatio* to a calling to a particular kind of religious life. He came to hold two interrelated beliefs about Christian vocation: (1) *all* Christians (not only monks) have a vocation, and (2) *every type of work* performed by Christians (not only religious activity) can be a vocation. Instead of interpreting *vocatio* as a call of a select group within the larger Christian fellowship to a special kind of life, Luther spoke of the double vocation of every Christian: spiritual vocation (*vocatio spiritualis*) and external vocation (*vocatio externa*). Spiritual vocation is God's call to enter the kingdom of God, and it comes to a person through the proclamation of the Gospel. This call is common to all Christians and is for all Christians the same ("*communis et similis*").[65] External vocation is God's call to serve God and one's fellow human beings in the world. It comes to a person through her station in life or profession (*Stand*).⁶⁶ This call, too, is addressed to all Christians, but to each one in a different way, depending on his particular station or profession ("*macht ein unterscheid*").⁶⁷

In *Kirchenpostille 1522*—a work in which Luther uses "vocation" for the first time as a *terminus technicus* "for a purely secular activity"⁶⁸—Luther gives an explanation of external vocation while answering the question of someone who feels without a vocation: "What if I am not called? What should I do? Answer: How can it be that you are not called? You are certainly in a station (*Stand*), you are either a husband or a wife, son or daughter, male or female servant."⁶⁹ To be a husband, wife, child, or servant *means to be called by God* to a particular kind of activity, it means to have a vocation. When God's spiritual call through the proclamation of the gospel reaches a person in her station or profession, it transforms these into a vocation. The duties of the station become commandments of God to her. In this way, Luther links the daily work of every Christian inseparably with the center of Christian faith: for a Christian, work in every profession, and not only in ecclesiastical professions, rests on a divine calling.

Two important and related consequences follow from Luther's notion of vocation. These insights make up the *novum* of Luther's approach to human work. First, Luther's notion of vocation ascribed much greater value to work than was previously the case. As Weber rightly observed, Luther valued "the fulfillment of duty in worldly affairs as the highest form which the moral activity of the individual could assume. . . . The only way of living acceptably to God was not to surpass worldly morality in monastic asceticism, but solely through the fulfillment of the obligations imposed upon the individual by his position in the world."⁷⁰ Second, Luther's notion of vocation *overcame the medieval hierarchy between* vita activa *and* vita contemplativa.⁷¹ Since every vocation rests on God's commission, every vocation is fundamentally of the same value before God.

Limits of the Vocational Understanding of Work

A responsible theology of work should seek to preserve Luther's insight into God's call to everyday work with its two consequences.

The way Luther (and especially later Lutheranism) developed and applied this basic insight is, however, problematic. Luther's notion of vocation has serious limitations, both in terms of its applicability to modern work, and in its theological persuasiveness.

Critique of Vocation

(1) Luther's understanding of work as vocation is *indifferent toward alienation* in work. In his view, two indispensable features sufficiently qualify a particular work theologically as vocation. The two features are the call of God and one's service to fellow human beings. The origin and purpose of work, not the inherent quality of work, define vocation. Hence it seems that virtually *every* type of work can be a vocation, no matter how dehumanizing it might be (provided that in doing the work one does not transgress the commandments of God).[72] Although it could never be one's vocation to be a prostitute because it entails breaking God's commandment, the vocational understanding of work does not in any way prevent mindless work on the assembly line at a galloping pace from being considered as a vocation. Such broad applicability might seem a desirable feature for an understanding of work, especially since (as Calvin pointed out) it can give "singular consolation" to people whose work is "sordid and base."[73] But one can have broad applicability and the benefits of consolation only at the expense of the transforming potential for overcoming alienation in situations when transformation is both necessary and possible. If even the "lifting of a single straw" is a "completely divine"[74] work, there is no reason why the same description could not apply to the most degrading types of work in industrial and information societies.

(2) There is a *dangerous ambiguity* in Luther's notion of vocation. In his view, spiritual calling comes through the proclamation of the gospel, while external calling comes through one's station (*Stand*). It has proven difficult for Lutheran theology to reconcile the two callings in the life of an individual Christian when a conflict arises between them. "The history of Lutheranism as well as Lutheran ethics shows that Luther's bold identification of vocation [i.e., *vocatio externa*] with the call [i.e., *vocatio spiritualis*] led again and again to the integration of the call into vocation and vocation into occupation, and thus to the consecration of the *vocational-occupa-*

tional structure. 'Vocation began to gain the upper hand over the call; the Word of God on the right (gospel) was absorbed by the word of God on the left (law).'"[75]

(3) The understanding of work as vocation is easily *misused ideologically*. As already indicated, Luther elevated work in every profession to the level of divine service.[76] The problem arises when one combines such a high valuation of work with both indifference to alienation and the identification of calling with occupation. Since the notion of vocation suggests that *every* employment is a place of service to God[77]—even when human activity in work is reduced to "soulless movement"—this notion functions simply to ennoble dehumanizing work in a situation where the quality of work should be improved through structural or other kinds of change. The vocational understanding of work provides no resources to foster such change.

(4) The notion of vocation is not applicable to the increasingly mobile industrial and information society. Most people in these societies do not keep a single job or employment for a lifetime, but often switch from one job to another in the course of their active life. The half-life of most job skills is dropping all the time, so they have to change jobs. And even if they could keep their jobs, they often feel that being tied down to a job is a denial of their freedom and of the opportunity for development. Industrial and information societies are characterized by a *diachronic plurality of employments or jobs* for their members. Luther's understanding of external vocation corresponds necessarily to the singleness and permanence of spiritual calling. As there is one irrevocable spiritual calling, so there must be one irrevocable external calling.

Given Luther's affirmation of the singleness and static nature of external vocation, it is easy to understand why he regularly relates his comments about external vocation to a conservative interpretation of the body of Christ and adds the injunction: "Let each one remain in his vocation, and live content with his gift."[78] The injunction to "remain" and "be satisfied" is a logical consequence of the notion of vocation.[79] To change one's employment is to fail to remain faithful to God's initial commandment. The only way to interpret change of employment positively and at the same time hold to the notion of vocation is to assume a diachronic plurality of external vocations. The soteriological meaning of vocation, which serves as a paradigm for the socioethical understanding of vocation, however, makes such an assumption anomalous. For singularity and permanence are constitutive characteristics of the soteriological understanding of vocation.

(5) In industrial and information societies people increasingly take on more than one job or employment at the same time. *Synchronic plurality of employments or jobs* is an important feature of these societies. In Lutheran theology, *vocatio externa* as a rule refers to a single employment or job, which people hold throughout their lives. This corresponds, of course, to the singularity of *vocatio spiritualis*. Unlike much of Lutheran theology, Luther himself maintained that, since a person mostly belonged to more than one *Stand* (she might have been daughter, mistress, and wife, all at the same time), a person had more than one external vocation.[80] His sense of reality led him to break loose from the exegetical and dogmatic framework set up with the concept of vocation. He is more consistent with this concept when he exhorts a person not to "meddle" in another's vocation.[81] Strictly speaking, one may take work to be *vocatio* only if one assumes that a Christian should have just one employment or job.[82]

(6) As the nature of human work changed in the course of industrialization, vocation was reduced to gainful employment. Lutheran social ethic followed this sociological development and, departing from Luther but in analogy to the singularity of the *vocatio spiritualis*, reduced its notion of vocation to gainful employment.[83] The reduction of vocation to employment, coupled with the belief that vocation is the primary service ordinary people render to God, contributed to the modern fateful elevation of work to the status of religion. The religious pursuit of work plays havoc with the working individual, his fellow human beings, and nature.

Reinterpretation of Vocation?

In responding to these criticisms, one might be tempted to reinterpret the understanding of work as vocation in order to free it from theological inadequacies and make it more applicable to industrial and information societies. There are, however, both exegetical and theological arguments against doing so.

(1) Exegetes agree that Luther misinterpreted 1 Corinthians 7:20, the main proof text for his understanding of work. "*Calling* in this

verse is not calling *with* which, to which, or by which a man is called, but refers to the state in which he is *when* he is called to become a Christian."[84] Except in 1 Corinthians 7:20 (and possibly 1 Cor. 1:26), Paul and others who share his tradition use the term *klēsis* as a *terminus technicus* for "becoming a Christian." As 1 Peter 2:9 shows, *klēsis* encompasses both the call of God out of "darkness into his wonderful light" that constitutes Christians as Christians, and the call to conduct corresponding to this "light" (see 1 Pet. 1:15), which should characterize life of Christians.[85] Thus, when *klēsis* refers not to becoming a Christian but to living as a Christian, it does not designate a calling peculiar to every Christian and distinguishing one Christian from another, as Luther claimed of *vocatio externa*. Instead, it refers to the quality of life that should characterize *all Christians as Christians*.

(2) Theologically it makes sense to understand work as *vocatio externa* only if one can conceive of this *vocatio* in analogy to *vocatio spiritualis*. One has to start with the singularity and permanence of *vocatio spiritualis*, which individualizes and concretizes itself in the process of human response in the form of a *singular and permanent* vocatio externa. Even Luther himself, in a social ethic designed for a comparatively static society, could not maintain this correspondence consistently. One could weaken the correspondence between *vocatio spiritualis* and *vocatio externa* and maintain that when the one call of God, addressing all people to become Christians, reaches each individual, it branches out into a plurality of callings for particular tasks.[86] I do not find it helpful, however, to deviate in this way from the New Testament and from a dogmatic soteriological use of *vocatio*, especially since the New Testament has a carefully chosen term—actually a *terminus technicus*—to denote the multiple callings of every Christian to particular tasks both inside and outside the Christian church. I refer to the term *charisma*.

I propose that a theology of charisms supplies a stable foundation on which we can erect a theology of work that is both faithful to the divine revelation and relevant to the modern world of work. In the following pages I will first give a theological reflection on the Pauline notion of *charisma*, and second apply it to a Christian understanding of work, while developing further the theology of charisms as the application demands.

A Theological Reflection on Charisms

In recent decades the subject of charisms has been the focus of lively discussion, both exegetical and theological. As I argue here briefly for a particular understanding of charisms, my purpose is not merely to analyze Paul's statements but to develop theologically some crucial aspects of his understanding of charisms, and in this way set up a backdrop for a theology of work.

(1) One should not define *charisma* so broadly as to make the term encompass the whole sphere of Christian ethical activity. E. Käsemann has argued that the whole ethical existence of the Christian, the *nova obœdientia*, is charismatic.[87] No doubt, the whole new life of a Christian must be viewed pneumatologically, but the question is whether it is legitimate to describe it more specifically as *charismatic*. I cannot argue for this point within the confines of a book on work,[88] but must simply assert that it seems to me more adequate to differentiate, with Paul, between the *gifts and the fruit* of the Spirit. The fruit of the Spirit designates the general character of Christian existence, "the lifestyle of those who are indwelled and energized by the Spirit."[89] The gifts of the Spirit are related to the specific tasks or functions to which God calls and fits each Christian.

(2) One should not define *charisma* so narrowly as to include in the term only ecclesiastical activities. One interpretation limits the sphere of operation of charisms to the Christian fellowship, insisting that one cannot understand "charismatically the various activities of Christians in relation to their non-Christian neighbors."[90] But, using individual charisms as examples, it would not be difficult to show the impossibility of consistently limiting the operation of charisms to the Christian church. The whole purpose of the gift of an evangelist (see Eph 4:11), for instance, is to relate the gospel to *non*-Christians. To take another example, it would be artificial to understand contributing to the needs of the destitute (see Rom 12:8) as *charisma* when exercised in relation to Christians but as simple benevolence when exercised in relation to non-Christians. As the firstfruits of salvation, the Spirit of Christ is not only active in the Christian fellowship but also desires to make an impact *on the world through the fellowship*.[91] All functions of the fellowship—whether directed inward to the Christian community or outward to the world—are the result of the operation of the Spirit of God and are thus charismatic.

The place of operation does not define charisms, but the manifestation of the Spirit for the divinely ordained purpose.

(3) Charisms are not the possession of an elite group within the Christian fellowship. New Testament passages that deal with charisms consistently emphasize that charisms "are found throughout the Church rather than being restricted to a particular group of people."[92] In the Christian fellowship as the Body of Christ there are no members without a function and hence also no members without a *charisma*. The Spirit, who is poured out upon all flesh (Acts 2:17ff.), imparts also charisms to all flesh: they are gifts given to the Christian community irrespective of the existing distinctions or conditions within it.[93]

(4) The tendency to restrict charisms to an elite group within the Christian fellowship goes hand in hand with the tendency to ascribe an elite character to charisms. In widespread pneumatologies in which the Spirit's function is to negate, even destroy the worldly nature,[94] "charismatic" is very frequently taken to mean "extraordinary." Ecclesiologically we come across this restricted understanding of charisms in some Pentecostal (or "charismatic") churches that identify charismatic with the spectacularly miraculous.[95] A secular version of this "supernaturalistic reduction" confronts us in the commonly accepted Weberian understanding of *charisma* as an extraordinary quality of leadership that appeals to nonrational motives.[96] One of the main points of the Pauline theology of charisms is the overcoming of such a restrictive concentration on the miraculous and extraordinary. For this reason it is of great importance to keep the term *charisma* as a generic term for both the spectacular and the ordinary.[97]

(5) Traditional view of the impartation of charisms can be described as the addition model: "the Spirit joins himself, as it were, to the person, giving 'something' new, a new power, new qualities."[98] It might, however, be better to understand the impartation of charisms according to the interaction model:[99] a person who is shaped by her genetic heritage and social interaction faces the challenge of a new situation as she lives in the presence of God and learns to respond to it in a new way. This is what it means to acquire a new spiritual gift. No substance or quality has been added to her, but a more or less permanent skill has been learned.

WORK, SPIRIT, AND NEW CREATION   113

We can determine the relationship between calling and charisma in the following way: the general calling to enter the kingdom of God and to live in accordance with this kingdom that comes to a person through the preaching of the gospel becomes for the believer a call to bear the fruit of the Spirit, which should characterize all Christians, and, as they are placed in various situations, the calling to live in accordance with the kingdom branches out in the multiple gifts of the Spirit to each individual.

Work in the Spirit

But is there a connection between charismata and the mundane work? If there is, can a theology of work be based on a theology of charismata? And if it could, would such a theology of work have any advantages over the vocational understanding of work so that we could with good conscience leave the second in favor of the first? Can it be applied to work of non-Christians or is it a theology of work only for a Christian subculture? Does not a pneumatological understanding of work amount to theological ideology of human achievement? To these questions I now turn.

Theological Basis

If we must understand every specific function and task of a Christian in the church and in the world charismatically, then everyday work cannot be an exception. The Spirit of God calls, endows, and empowers Christians to work in their various vocations. The charismatic nature of all Christian activity is the *theological* basis for a pneumatological understanding of work.

There are also some *biblical* references that can be taken to suggest a pneumatological understanding of work. We read in the Old Testament that the Spirit of God inspired craftsmen and artists who designed, constructed, and adorned the tabernacle and the temple. "See, the Lord has chosen Bezalel . . . and he has filled him with the Spirit of God, with skill, ability and knowledge in all kinds of crafts . . . and . . . the ability to teach others" (Exod. 35:2-3). "Then David gave his son Solomon . . . the plans of all that the Spirit had put in his mind for the courts of the temple of the Lord" (1 Chron. 28:11-12). Furthermore, judges and kings in Israel are often said to do their tasks under the anointing of the Spirit of God (see Judg. 3:10; 1 Sam. 16:13; 23:2; Prov. 16:10).[100]

As they stand, these biblical affirmations of the charismatic nature of human activity cannot serve as the basis for a pneumatological understanding of *all* work, for they set apart people gifted by the Spirit for various extraordinary tasks from others who do ordinary work. But we can read these passages from the perspective of the new covenant in which *all* God's people are gifted and called to various tasks by the Spirit. In this case they provide biblical illustrations for a charismatic understanding of the basic types of human work: intellectual (e.g. teaching) or manual (e.g. crafts) work, *poiesis* (e.g. arts and crafts) or *praxis* (e.g. ruling). All human work, however complicated or simple, is made possible by the operation of the Spirit of God in the working person; and all work whose nature and results reflect the values of the new creation is accomplished under the instruction and inspiration of the Spirit of God (see Isa. 28:24-29).

Work as Cooperation with God

If Christian mundane work is work in the Spirit, then it must be understood as *cooperation with God*. *Charisma* is not just a call by which God bids us to perform a particular task, but is also an inspiration and a gifting to accomplish the task. Even when *charisma* is exercised by using the so-called natural capabilities, it would be incorrect to say that a person is "enabled" irrespective of God's relation to him. Rather, the enabling depends on the presence and activity of the Spirit. It is impossible to separate the gift of the Spirit from the enabling power of the Spirit.[101] When people work exhibiting the values of the new creation (as expressed in what Paul calls the "fruit of the Spirit") then the Spirit works in them and through them.

The understanding of work as cooperation with God is implied in the New Testament view of Christian life in general. Putting forward his own Christian experience as a paradigm of Christian life, Paul said: "it is no longer I who live, but Christ who lives in me; and the life I now life in the flesh I live by faith in the Son of God" (Gal. 2:20). That Paul can in the same breath make such seemingly contradictory statements about the acting agent of Christian life ("I no longer live, *Christ lives* in me" and "*I live* my life in the flesh") testifies unmistakably that the whole Christian life is a life of cooperation with God through the presence of the Spirit. A Christian's mundane work is no exception. Here, too, one must say: I work, and the Spirit of the resurrected Christ works through me.

Since the Spirit who imparts gifts and acts through them is "a guarantee" (2 Cor. 1:22; cf. Rom. 8:23) of the realization of the eschatological new creation, cooperation with God in work is proleptic cooperation with God in God's eschatological *transformatio mundi*. As the glorified Lord, Jesus Christ is "present in his gifts and in the services that both manifest these gifts and are made possible by them."[102] Although his reign is still contested by the power of evil, he is realizing through those gifts his rule of love in the world. As Christians do their mundane work, the Spirit enables them to cooperate with God in the kingdom of God that "completes creation and renews heaven and earth."[103]

A Pneumatological Approach to Work: Does It Solve Anything?

In the last two chapters I will develop some of the most important aspects of a pneumatological understanding of work. Here I want to show that this understanding of work is not weighed down by the serious deficiencies of the vocational understanding of work.

(1) The pneumatological understanding of work is free from the portentous ambiguity in Luther's concept of vocation, which consists in the undefined relation between spiritual calling through the gospel and external calling through one's station. The resurrected Lord alone through the Spirit calls and equips a worker for a particular task in the world. Of course, neither the Spirit's calling nor equipping occur in a social and natural vacuum; they do not come, so to speak, directly from Christ's immaterial Spirit to the isolated human soul. They are mediated through each person's social interrelations and psychosomatic constitution. These mediations themselves result from the interaction of human beings with the Spirit of God. Yet charisms *remain different from their mediations* and should not be reduced to or confused with them.[104] For the Spirit who gives gifts "as he wills" (1 Cor. 12:11) *by* social and natural mediation is not the Spirit *of* human social structures or of a persons' psychosomatic makeup, but the Spirit of the crucified and resurrected Christ, the firstfruits of the new creation.

(2) The pneumatological understanding of work is *not as open to ideological misuse* as the vocational understanding of work.[105] It does not proclaim work meaningful without simultaneously attempting to humanize it. Elevating work to cooperation with God in the pneumatological understanding of work implies an obligation to overcome alienation because the individual gifts of the person need to be taken seriously. The point is not simply to interpret work religiously as cooperation with God and thereby glorify it ideologically, but to transform work into a charismatic cooperation with God on the "project" of the new creation.

(3) The pneumatological understanding of work is easily applicable to the increasing *diachronic plurality* of employments or jobs that characterize industrial and information societies. Unlike Christian calling, *charisma*—in the technical sense—is not "irrevocable" (see Rom. 11:29). True, a person cannot simply pick and choose her *charisma*, for the sovereign Spirit of God imparts charisms "as he wills" (1 Cor. 12:11). But the sovereignty of the Spirit does not prohibit a person from "earnestly desiring" spiritual gifts (1 Cor. 12:31; 14:1,12) and receiving various gifts at different times.[106] Paul presupposes both a diachronic and a synchronic plurality of charisms.

The diachronic plurality of charisms fits the diachronic plurality of employment or jobs in modern societies. Unlike in the vocational understanding of work, in the pneumatological understanding of work one need not insist that the occupational choice be a single event and that there be a single right job for everyone[107] (either because God has called a person to one job or because every person possesses a relatively stable pattern of occupational traits). People are freed for several consecutive careers in rapidly changing work environments; their occupational decisions need not be irrevocable commitments but can be repeatedly made in a continuous dialogue between their preferences and talents on the one hand, and the existing job opportunities on the other.[108]

In any case, one can change jobs without coming under suspicion of unfaithfulness. If the change is in harmony with the *charisma* given, then changing can actually be an expression of faithfulness to God, who gave the *charisma* and readiness to serve fellow human beings in a new way. There is no need to worry that in the absence of a permanent calling, human life will be "turned topsy-turvy"[109] (as Calvin thought) or that human beings will "spend more time in idleness than at work"[110] (as the Puritans feared). Rather, freedom from the rigidity of a single, permanent vocation might season with creativity and interrupt with rest the monotonous lives of modern workaholics.

(4) It is also easy to apply the pneumatological understanding of work to the *synchronic plurality* of jobs or employments. In Paul's view every Christian can have more than one *charisma* at any given time. His aim is that Christians "excel in gifts" (1 Cor. 14:12), provided they exercise them in interdependence within the community and out of concern for the common good. The pneumatological understanding of work frees us from the limitation of being able to theologically interpret only a single employment of a Christian (or from the limitation of having to resort to a different theological interpretation for jobs that are not primary). In accordance with the plurality of charisms, there can be a plurality of employments or jobs without any one of them being regarded theologically as inferior, a mere "job on the side." The pneumatological understanding of work is thus also open to a redefinition of work, which today's industrial and information societies need.[111]

Spirit and Work in *Regnum Naturae*

As I have sketched it, the pneumatological understanding of work is clearly a theology of *Christian* work. The significance and meaning of Christians' work lie in their cooperation with God in the anticipation of the eschatological *transformatio mundi*. The power enabling their work and determining its nature is the Holy Spirit given when they responded in faith to the call of God in Christ.

But what about the work of non-Christians? Traditionally theologians simply bypassed the issue as uninteresting. Although Luther, for instance, did not apply the concept of vocation to the work of non-Christians,[112] he reflected little in his writings on the theological significance of their work. This is understandable, given the identity of church and society in the *Corpus Christianum* that Luther and other seminal theologians of the past presupposed. In much of the world throughout history, however, church and society were never identified, and the cradle of the *Corpus Christianorum* is becoming its grave: in the Western world a clear and irretrievable separation between church and society is taking place. Since Christians today live in religiously pluralistic societies, their theologies of work must incorporate reflection on the work of non-Christians. Hence my next step is to indicate the implications of a pneumatological theology of work for understanding non-Christians' work.

What is the relation of the work of non-Christians to the new creation? The answer to this question is implicit in the way I have determined the relation between the present and the future orders. If the world will be transformed, then the work of non-Christians has in principle the same ultimate significance as the work of Christians: insofar as the results of non-Christians' work pass through the purifying judgment of God, they, too, will contribute to the future new creation. In Revelation one reads that the kings of the earth and the nations will bring their splendor, glory, and honor into the New Jerusalem (Rev. 21:24, 26). It makes perhaps the best sense to take this enigmatic statement to mean that all pure and noble achievements of non-Christians will be incorporated in the new creation.

But is it possible to understand the work of non-Christians *pneumatologically*? Charisms are specifically ecclesiastical phenomena. They are gifts given to those who acknowledge Jesus as Lord. How, then, can anything we learned about the nature of work from the theology of *charisms* apply to the work of non-Christians? The answer depends on how we conceive of the relationship between the Spirit of God and the non-Christians. I can only sketch an approach to this extremely complex and not sufficiently investigated subject here.

First, if we affirm that Christ is the Lord of all humanity—indeed, of the whole universe—and not only of those who profess him as their Lord, and that he rules through the power of the Spirit, then we must also assume that the Spirit of God is active in some way in all people, not only in those who consciously live in the Spirit's life-giving power. As Basil of Caesarea observes in his *De Spiritu Sancto*, creation possesses nothing—no power, no motivation, or ingenuity needed for work—that it did not receive from the Spirit of God.[113] There is hence an important sense in which all human work is done "in the power of the Spirit."

Second, one and the same Spirit of God is active both in the Church and in the world of culture. As the firstfruits of the new creation, the Spirit is active in the Church, redeeming and sanctifying the people of God. In the world of culture the Spirit is active sustaining and developing humanity. The difference in the activity of the Spirit in these two realms lies not so much in the different purposes of the Spirit with the two groups of human beings, as in the nature of the receptivity of human beings. Third, the goal of the Holy Spirit in the church and in the world is the same: the Spirit strives to lead both the realm of nature (*regnum naturæ*) and the realm of grace (*regnum gratiæ*) toward their final glorification in the new creation (*regnum gloriæ*).[114]

Since in the realm of grace the Spirit is active as the firstfruits of the coming glory, which is the goal of the realm of nature, we must think of the Spirit's activity in the realm of nature as analogous to its activity in the realm of grace. What can be said of the work of Christians on the basis of the biblical understanding of charisms can also be said by analogy of the work of non-Christians. Revelation of the future glory in the realm of grace is the measure by which events in the realm of nature must be judged. To the extent that non-Christians are open to the prompting of the Spirit, their work, too, is the cooperation with God in anticipation of the eschatological transformation of the world, even though they may not be aware of it.

A CHRISTIAN IDEOLOGY OF WORK?

Work as cooperation with God in the eschatological transformation of the world! Work in the Spirit! These are lofty words about human work. But is it not true that work reflects not only the glory of human cooperation with God but also the misery of human rebellion against God? This is, indeed, a testimony of Genesis 2 through 3, which explains how pleasant work in a garden (2:15) became futile toil outside of it (3:17ff.). The experience of most working people confirms it. The statement Wolterstorff makes about art is *a forteriori* true of work: it "reeks of murder, and oppression, and enslavement, and nationalism, and idolatry, and racism, and sexism."[115]

Given the drudgery of much of modern work, the exploitation of workers, and the destruction of nature through human work, does not the talk about working in the Spirit and about the eschatological significance of work sound suspect? Does it not amount to a glorification of work that conceals the debasement of workers? Is a theology of work only an ideology of work in disguise?

God's Judgment of Human Work

The understanding of work as cooperation with God in the *transformatio mundi* is not a general theory of all human work. It is not applicable to every type of work and to every way of working, for the simple reason that the new creation will not incorporate everything found in the present creation. When God creates a new world he will not indiscriminately affirm the present world. Such promiscuous affirmation would be the cheapest of all graces, and hence no grace at all. The realization of the new creation cannot bypass the Judgment Day, a day of negation of all that is negative in the present creation.[116]

Paul's reflection on the ultimate significance of missionary work in the face of God's judgment (1 Cor. 3:12-15) might give us a clue to understanding God's judgment in relation to human work in general. Like the test of fire, God's judgment will bring to light the work that has ultimate significance since it was done in cooperation with God. Like gold, silver, and precious stones (see 1 Cor. 3:12), such work will survive the fire purified. But the Judgment Day will also plainly reveal the work that was ultimately insignificant because it was done in cooperation, not with God, but with the demonic powers that scheme to ruin God's good creation. Like wood, hay, and straw, such work will burn up, for "nothing that is impure will ever enter" the New Jerusalem (Rev. 21:27). Every understanding of work as cooperation with God that does not include the theme of judgment is inadequate. As we have to pattern our work according to the values of the new creation, so we also have to criticize it in the light of the eschatological judgment.

In relation to God's judgment on human work, it is important to distinguish between what might be called the moral and the ontological value of human work. I have already argued against ascribing eschatological significance merely to the attitude of love exhibited in work.[117] It would also be insufficient to attach eschatological significance only to the results of work done in love.[118] "Man's envy of his neighbor" (Eccles. 4:4), as the realistic ecclesiast puts it, spurs him on to many of the best human achievements. Do they lose their inherent value because they were done out of ethically impure motives? Every noble result of human work is ultimately significant. It is possible that the fire of judgment will not only burn up the results of work, the worker herself escaping "the flames" (1 Cor. 3:15),[119] but that the flames of "the absolutely searching and penetrating love of God"[120] will envelop the evil worker while her work is purified and preserved.

The reality of Judgment makes it clear that relating human work positively to God's new creation does not amount to an ideological glorification of work. It lies in the affirmation that the work has meaning in spite of the transitoriness of the world. If human work is in fact "chasing after wind" (Eccles. 4:4)—whether or not one experiences it subjectively as meaningful—it is not so because of the transitoriness of the world, but because of the evilness of the work. All work that contradicts the new creation is meaningless; all work that corresponds to the new creation is ultimately meaningful. This should serve as an encouragement to all those "good workers" who see themselves in the tragic figure of Sisyphus. In spite of all appearances, their work is not just rolling a heavy rock up a hill in this earthly Hades; they are preparing building blocks for the glorified new creation. Furthermore, all those weighed down by the toil that accompanies most of human work can rest assured that their sufferings "are not worth comparing with the glory" of God's new creation they are contributing to (Rom. 8:18).

Work Against the Spirit

What is the relationship between the Spirit of God and the work that deserves God's judgment? There is a sense in which all human work is done in the power of the Spirit. The Spirit is the giver of all life, and hence all work, as an expression of human life, draws its energy out of the fullness of divine Spirit's energy. When human beings work, they work only because God's Spirit has given them power and talents to work. To express the same thought in more traditional terminology, without God's constant preserving and sustaining grace, no work would be possible.

But a person can misuse his gifts and exercise them against God's will. Through his work he can destroy either human or natural life and hence contradict the reality of the new creation, which preserves the old creation in transfigured form. The circumstance that the gifts and energies that the Spirit gives can be used against the will of the Spirit results from the Spirit's condescension in history: by giving life to the creation, the Spirit imparts to the creation the power for independence from the Spirit's prompting. Because the Spirit creates human beings as free agents, work in the power of the Spirit can be done not only in accordance with but also in contradiction to the will of the Spirit; it can be performed not only in cooperation with the Holy Spirit who transforms the creation in anticipation of the glorious new creation, but also in collaboration with that Unholy Spirit who strives to ravage it.

CHAPTER 5

Work, Human Beings, and Nature

The basic contention of the previous chapter and the main thesis of this book is that Christians should understand their mundane work as "work in the Spirit": the Spirit of God calls and gifts people to work in active anticipation of the eschatological transformation of the world. But what does that mean concretely? What view of work and of human beings do we get when we apply Pauline teaching on charisms to mundane work? From a pneumatological perspective, how should we conceive of the relationship of work to leisure, to human needs, and to nature? How does this view of work deal with the problem of alienation and the humanization of work? In the remainder of the book I will attempt to answer these questions.

A comprehensive theology of work would need to discuss these issues much more exhaustively than I am able to do here. If I were to attempt to develop a full-blown theology of work, I would far exceed the limits of this book. What I intend to do here is only to sketch some basic aspects of work's relation to human beings—to their nature, their needs, and their other significant activities—and to their natural environment.[1] I will deal first with the question of the centrality of work in the Christian life and of the anthropological significance of work. Second, I will discuss the problem of leisure (particularly of worship as its central aspect) and its relation to the world of work. Then I will deal with the relation of work to the endangered natural world, which is the object or environment of work. Finally, I will address the question of dynamic and expanding human needs whose satisfaction is the conscious or unconscious purpose of human work. In the next chapter, to illustrate what a comprehensive theology of work could look like, I will more extensively discuss from a pneumatological perspective one of the key problems facing work today: the problem of alienation and humanization.

To discuss the anthropological significance of work, the relationship between work and leisure, work and nature, and work and human needs, means to look at the single reality of human work from various perspectives. In order to do the topic justice, one would have to address all these topics from the perspective of each of them. But in that case, the discussion would have to circle around its subject, and a good deal of repetition would be unavoidable. I have settled for a compromise: I will shun repetition as much as possible but set up the signposts pointing out the connecting paths between different sections. I will leave to the reader the simple task of carrying the needed materials from one section to the other.

SPIRIT, WORK, AND HUMAN BEINGS

Work—A Central Aspect of Christian Life?

When God calls people to become children of God, the Spirit gives them callings, talents, and "enablings" (charisms) so that they can do God's will in the Christian fellowship and in the world in anticipation of God's eschatological new creation. All Christians have several gifts of the Spirit. Since most of these gifts can be exercised only through work, work must be considered a central aspect of Christian living. Anthropological reflection will corroborate this pneumatological insight.[2]

One consequence of the centrality of work in the Christian life and of the fact that every work of a Christian should be done under inspiration of the Spirit is that there is no hierarchical valuation of the various tasks a Christian may perform. Every task—be it "spiritual or physical,"[3] as Luther said—has fundamentally the same dignity.

Overvaluation of Work?

Mundane work as proleptic cooperation with God's Spirit in God's eschatological transformation of the world! Is this view of work not merely a Christian mirroring of the modern fascination with work? Is not, in fact, this view of work so high that it dwarfs both the modern religion of work and the Protestant elevation of work to divine service that contributed to it?

These questions become even more pressing if we keep in mind the nature of the motivation to work in a pneumatological theology of work. In the vocational understanding of work, God addresses human beings, calling them to work, and they respond to God's call primarily by obedience. They work out of a sense of *duty*. In a pneumatological understanding of work, God does not first and foremost command human beings to work, but empowers and gifts them for work. They work, not primarily because it is their duty to work, but because they experience the inspiration and enabling of God's Spirit and can do the will of God "from the heart" (Eph. 6:6; cf. Col. 3:23). When a person does her secular tasks in grateful obedience for the new life God has given her, she also works out of the experience of God's grace; but grace remains, so to speak, in the background. Grace only "compels" her to act (see 2 Cor. 5:14).[4] But when grace gifts and enables a person to do a particular task, then it stands at the very heart of her work. The appropriate response to such an experience of grace is not so much naked (though thankful!) obedience as it is joyful willingness to employ the capabilities conferred in the entrusted "project." Though not fully absent, the sense of duty gives way to the sense of inspiration.

This altered motivation for work better suits modern society, in which knowledge is becoming a main resource and hence freedom and creativity indispensable features of work. But is not the price that a pneumatological understanding of work has to pay for this achievement too high? For it seems that it contributes to the modern fascination with work instead of relativizing it. The section in this chapter entitled "Spirit, Work, and Leisure"[5] puts forward part of the answer to this question. There I show that worship is the fundamental aspect of human existence that limits work. Here I want to investigate what value Christian theology should ascribe to work by investigating the anthropological significance of work. I will start with a brief discussion of some ancient Greek and modern views.

Work: Curse and Religion

Influential thinkers of ancient Greece considered work a necessary evil, without any value in and of itself. Just as the only purpose of war is peace, thought Aristotle, so the only purpose of work is leisure.[6] One reason for such a negative attitude toward work was the conviction that labor, which is necessitated by human bodily needs, is slavish. "To labor meant to be enslaved by necessity and this enslavement was inherent in the conditions of human life."[7] The other reason was the persuasion that labor caused degeneration of both the human body and soul because it robbed people of the leisure necessary for physical, intellectual, and moral health. Summarizing the detrimental effects of labor, Plato speaks of workers "whose souls are bowed and mutilated by their vulgar occupations even as their bodies are marred by their mechanical arts."[8] Given such views of the nature and effects of work, it is not surprising to find Hesiod contrasting his own unfortunate generation, in which people "never rest from labour,"[9] with the first generation of human beings, the "golden race of men," who lived "like gods . . . remote and free from toil" because the fruitful "earth unforced bore them fruit abundantly and without stint."[10] The goal of human beings is to live like gods, free from the evil of necessary work.

In modern times attitudes toward work have changed radically. Many modern philosophers were fascinated with the achievements of human work, but it was Thomas Carlyle who sang its highest praises. With him, work took on explicitly religious overtones. Carlyle expressed his view of the deepest significance of work most aptly when he altered the old monastic rule *ora et labora* (pray and work) into *laborare est orare* (working is praying). With Carlyle, mundane work replaced prayer to God and became a means of secular salvation. Work is "the latest Gospel in this world,"[11] wrote Carlyle, because it helps people find their true selves and elevates them "from the low places of this Earth, very literally, into divine Heavens."[12]

Though there would be very few people today who would praise work with the same enthusiasm as Carlyle did, still, the majority would see work more as a savior of the human race than as a corruptor of human nature. Should Christians follow suit? Against modern tendencies toward the divinization of work, biblical tradition soberly points to the drudgery and futility that accompany much of human work in a sinful and transitory world, even work that is done in the power of the Spirit. Against the ancient demonization of work, on the other hand, biblical tradition affirms work as a fundamental dimension of human existence.

Work: a Fundamental Dimension of Human Existence

As can be seen from both creation accounts in Genesis, the Old Testament considers work essential to human life.[13] One can debate the precise meaning of the *imago Dei* in the first creation account,[14] but there is no doubt that the creation of human beings in the image of God is closely related to work. For we read that God created human beings in his image *"in order to have dominion* over the fish of the sea . . ." (Gen. 1:26).[15] The text does not mention work explicitly, but since human beings can fulfill their God-given task only by working, it is obvious that this *locus classicus* of Christian anthropology considers work to be fundamental to human existence. Only as working beings—though not exclusively or even primarily as working beings—can human beings live in accordance with the intention of their Creator.

The second creation account specifically addresses the question of human work and emphasizes its anthropological significance equally strongly. The account starts with the statement that "there was no man to till the ground" (Gen. 2:5) and concludes with the statement that God sent the first human pair out of the garden of Eden to "till the ground from which they were taken" (Gen. 3:23). In this framework, the narrative describes how God created Adam and placed him in the garden of Eden to work in it, to "till it and keep it" (Gen. 2:15). At the end of the story we are told of the strenuousness of work and the precariousness of its results as a consequence of the Fall (Gen. 3:17ff.). How the paradisaical work of tilling and keeping became the exhausting and frustrating toil outside the garden of Eden is one important aspect of the drama of the narrative.[16] It is therefore correct to say that, according to

Genesis 2, "the work of human beings . . . belongs to the very purpose for which God originally made" them.[17] In that human beings "go forth to their work" (Psalm 104:23), they fulfill the original plan of the Creator for their lives. Reformers were quite right to liberate the working "Martha" from the dominance of the contemplative "Mary."[18]

Yet the Old Testament is not blind to the seamy side of human work. Though it is completely mistaken to talk about the "biblical curse of necessary work,"[19] many workers—even very good and satisfied workers—feel that Genesis 3:17-19 reflects their own experience as workers. There we read:

> . . . Cursed is the ground because of you;
> in toil you shall eat of it all the days of your life;
> thorns and thistles it shall bring forth to you;
> and you shall eat the plants of the field.
> In the sweat of your face
> you shall eat bread
> till you return to the ground . . .

There is no suggestion in this text of work itself being a curse. But it clearly says that, as a consequence of the curse against the ground, work has assumed the *character of toil*. The narrative explains the drudgery of human work in two ways. First, human beings may no longer eat the fruit of the "garden" obtained through the fulfilling work of cultivation (Gen. 2:15). They have to eat "the plants of the field," which they must till by the "sweat of their face." Hardship accompanies human work. Second, after the Fall, even strenuous work cannot prevent thorns and thistles from growing in the crop. Human work is "threatened by failures and wastes of time and often enough comes to nothing."[20]

Work and Human Nature

In a Christian anthropology the key question in relation to work is not how central to human existence work is, but what influence work has on human nature. The (actual or perceived) relation of work to human nature is also a key to understanding the modern fascination with work.

Some social analysts observe the captivation with work in modern, economically developed societies and conclude that the Protestant work ethic is still alive and well today. But there is very little that is either specifically Protestant (religious) or ethical about the contemporary drive to work. This is not to deny the historical contribution of the Protestant work ethic to modern workaholism. But after Western civilization has climbed up the ladder of the Protestant work ethic to a state in which incessant work has become one of its main features, it has pushed this ladder aside but continued to work even more frantically. Work thrives today more on the insatiable hunger for self-realization than on the Protestant work ethic. In their own eyes and in the eyes of their contemporaries,[21] modern human beings are what they do. The kind of work they do and what they accomplish or acquire through work provides a basic key to their identity. Hence the narcissistic and isolating preoccupation with "self" need not discourage hard work but can even stimulate it.[22] The contemporary religion of work has little to do either with worship of God or with God's demands on human life; it has much to do with "worship" of self and human demands on the self.[23] Hence the question of how a pneumatological understanding of work relates to the modern captivation with work turns (in part) into another question: how does this understanding of work relate to the central anthropological significance assigned to work in contemporary societies?

Work and Personality

As a recent public opinion poll in the United States shows, one of the top four characteristics of desirable work is that it provide the chance for workers to develop skills. Increasingly, people think that the work place should not only be a place where "profits thrive" but also where "people flourish." Even some companies are striving to ensure that "the total work package . . . be varied enough to provide mental stimulus for personal growth."[24] (This attempt has less to do with the benevolence of companies than with the fact that only people who flourish can be good performers in the economies of information societies, in which knowledge has become the main resource.)

The belief that people should develop through work rests on the insight that work can and does significantly influence human personality. As noted earlier, Adam Smith was among the first to translate the idea of moral development through work (an idea that was well rooted in the Christian tradition) into the categories of *anthropological* development through work.[25] Though he did not deny the existence of innate differences between human beings, Smith claimed that the differences in intelligence and skills between adult individuals are to a large extent the consequence of the kind of work they are doing. The idea of development through work was picked up in particular by Continental philosophers in the idealist tradition and became an integral part of their thinking about the nature and significance of work.[26]

Sociological investigations corroborate the anthropological thesis about human development through work, especially if one does not look at work simply as an isolated activity but as occupation or career. If human beings are to work at all, they have to internalize through the process of socialization a set of extra-individual conditions with which they must cope. This process forms their work personality, which is a significant sub-area of general personality.[27] Furthermore, jobs seem to contribute to major segments of people's identities—they influence awareness of who they are, where they have been, and where they can expect to go.[28]

At first glance it might seem that the idea of human personality's being shaped by work and of development through work might not easily fit into a pneumatological understanding of work. For charisms are not achievements of human beings and products of their environment, but gifts of God. One would misinterpret the nature of God's action, however, if one were to think of charisms as coming only, so to say, "vertically from above." Some capabilities human beings are born with, and many others they acquire in the course of their life through interaction with other human beings and their cultural and natural environments. These capabilities, too, are gifts of God's Spirit. On the basis of passages like Exodus 31:2f., Calvin stressed that all human skills stem from the operation of the Spirit of God.[29]

One should, moreover, not merely passively receive the gifts of the Spirit. The interaction model of the impartation of charisms[30] requires activity by the recipient, an attitude of active receptivity. For according to this model, charisms are partly constituted through the way a person relates to situations she encounters or lives in; she acquires new charisms or develops existing charisms partly through her own activity. It is the task of every Christian to seek new charisms (see 1 Cor. 14:12) and to reactivate and develop existing charisms (see 2 Tim. 1:6), whether during the work experience or outside of it.

In the New Testament, however, the reception, development, and use of gifts must be accompanied by the nurture of the fruit of the Spirit (Gal. 5:22f.). The fruit of the Spirit, which consists in the values of the new creation, determines how the gifts of the Spirit should be used. In a pneumatological understanding of work, the development of human beings through their work is, therefore, taken out of the domain of the individualistic search for self-actualization[31] and put in the context of concern for God's new creation. This notion of self-development is consciously and unabashedly value-laden. Any useful notion of self-development must be value-laden, since it supposes a normative understanding of human nature. If it were value-free, it would be tautological, because "any action of a human being is part of a pattern of actions that actualizes the self of that particular human being."[32]

How does my own development relate to God's new creation? Since I, too, hope to be part of God's new creation as a fulfilled and satisfied individual, care for the new creation includes care for my own development. My own development is an end in itself because it is integral to the new creation and hence a good to be affirmed. But my development is not self-contained because I as an individual am not self-contained. I can be fulfilled only when the whole creation has found its fulfillment, too. Because I am an essentially social and natural being, my development, which is an end in itself, is at the same time a means of benefitting others. Hence I cannot concentrate on the development of my capabilities while disregarding their use for the well-being of the social and natural world I inhabit. My development must be attuned to the well-being of the whole creation. Therefore, to look at work from the perspective of social and ecological practice and to look at it from the perspective of a worker's self-development are not mutually incompatible alternatives. Since individuals, society, and nature form an integrated unity, these perspectives on work are complementary.

Human Beings: Their Own Products?

Some influential philosophers found it insufficient to say that human beings develop through work. They made the stronger claim that human beings are the *result* of their work. Following the lead of Greek philosophy, Western theological and philosophical traditions for centuries understood the human being primarily as the *animal rationale:* it is rationality that distinguishes humans from other living beings. Much like Hegel before him (who conceived of a person as "a series of his actions"[33]) and Nietzsche after him (who claimed that "the doer" is merely a fiction added to the deed—the deed is everything"[34]), Marx broke radically with this tradition and defined human beings through their activity. But more than any other thinker he stressed the singular anthropological significance of human work. For Marx the human person is primarily *homo creator.*[35]

In Marx's thought, human work is anthropologically central in three related ways. First, historically, human beings begin to differ from animals at the moment they start producing their own means of sustenance.[36] Second, what human beings are at a particular point in their historical development is determined by the character of their material production. "What individuals are *coincides* with their production, both with what they produce and with how they produce."[37] Human beings are not static entities; their nature changes with the character of their work. As a species (not as individuals), they are therefore always and exclusively their "own product and result."[38] Third, Marx closely related his normative understanding of human nature with a particular kind of human work. In his view, the "whole character" of a species lies "in the character of its life activity."[39] The life activity of human beings is work as a free activity performed in mutual service. Hence they are most themselves—they are "at home," as Marx puts it—when they work freely for one another.

But is it theologically adequate to conceive of human beings as products of their own work? And are they "at home" the most when they work? While we must affirm that human beings can and do develop through work, we must firmly deny that as human beings they are constituted through work.[40] Neither as individuals nor as a species can human beings bestow upon themselves their humanity. Only God can do that. By relating to them as partners, God makes them into human beings.[41] If God's relation to human beings is the key to their humanness, then their communion with God is the key to their true identity. Human beings are truly themselves only when their relation to God is one of acceptance in faith and love of God's relation to them. They are truly themselves when they are in communion with their Creator and Redeemer through the Spirit.

Since the presence of the Spirit is the key to human identity, there is no fear that by not working human beings may lose themselves. In fact, they will lose themselves if in working they cling to themselves either in vanity or in self-deprecation (Mark 8:36). Any attempt to "give birth to oneself," so to speak, through work is to ask too much from work and will inevitably result in alienation. If, on the other hand, through the Spirit human beings start seeking God's new creation by worshiping God and cooperating with God in the world, then they will find themselves (see Matt. 6:33). In that the Spirit liberates human beings from slavery to their own selves and opens them for the reality of God's new creation, they become free, not only to work, but also to worship and to play.

SPIRIT, WORK, AND LEISURE

Social and scholarly interest in leisure is high today. But it is notoriously difficult to determine what leisure is. Everybody uses the word, observes Neulinger, "but hardly anyone can agree on what it means."[42] I will not enter the discussion of this problem here[43] but will simply propose a definition of leisure leaning on my definition of work. If one keeps in mind that work and leisure are polar concepts and assumes the definition of work that I have given in the Introduction,[44] then the following formal definition of leisure activity suggests itself: *Leisure is an activity that is primarily an end in itself and hence (as activity) satisfies a need of acting* individuals but is at the same time not necessary for, or done with the primary goal of, meeting other needs, either of the acting individuals themselves or of their fellow creatures.[45]

As I have defined them, work and leisure are polar but not mutually exclusive activities; the dividing line between them is not very sharp.[46] They are two clearly distinct spheres of human activity, since by definition there can never be a complete identity between the two. On the other hand, the two types of activities need not be completely separate. Work and leisure represent opposite ends of a continuum, so that under certain conditions work can have characteristics of leisure activities without ceasing to be work, and vice versa. And, though work can never become leisure, the more it resembles leisure, the more humane it will be.[47]

Since leisure activity is an end in itself, it is by definition an enjoyable activity. How much room is there for such enjoyable noninstrumental activities in industrial and information societies?

Dream of a Leisured Society?

Contrary to what many people think, industrialization initially brought a sharp increase in the average amount of time people spent working. Prior to the eighteenth-century, secular or religious holidays had taken up the better part of six months. In the first stages of industrialization many of these holidays were done away with, and workers were often forced to work sixteen hours a day or even more. In 1860, European workers still worked eighty hours a week on the average. In the later stages of industrialization, however, there was a steady, long-term decline in the industrial workweek brought about by pressure from workers, more rational organization of work, and new technological developments.

As technological development gained pace in the fifties and sixties, it seemed that humanity stood on the threshold of a new age of leisure. So "the 1967 testimony before a U.S. Senate subcommittee indicated that by 1985 people could be working just 22 hours a week or 27 weeks a year, or could retire at 38."[48] The promises of a microelectronic revolution seduced some sociologists and economists even to speculate that in the future one day a week of work might be more than enough for the production of all consumer goods.⁴⁹ The time seemed ripe to pronounce work dead and to call for leisure values to replace work values.⁵⁰

It is possible (even likely) that work productivity will grow faster than human needs for consumer goods, so that in some distant future the time needed for production of consumer goods will shrink to insignificance. But that does not mean that the amount of work done will be significantly reduced, let alone that work will be abolished. Judging by developments taking place presently, the very opposite might happen. For, in spite of increased efficiency through permanent technological innovations, according to a Harris survey, "the amount of leisure enjoyed by the average U.S. citizen has shrunk by 37% since 1973. Over the same stretch, the average work-week, including commuting time, has moved up from less than 41 hours to nearly 47."⁵¹ In contemporary technological civilization, which can boast of remarkable labor-saving innovations, human beings paradoxically work more than they have ever worked before.⁵²

Not only is work not dead; at least in some circles, work values are not dead either. True, lower-level employees often have a purely instrumental view of work. For them work has no inherent value. It functions as a necessary means of keeping soul and body together and of sustaining the pursuit of happiness outside work (although closer examination would, I suspect, show that the actual significance of work in the lives of lower-level employees by far exceeds its perceived significance). Since leisure provides a greater scope for self-determination than work, it is the main focus of their dreams of personal fulfillment.⁵³ But an increasing number of professionals, whose work allows a great measure of freedom, still find fulfillment in work, so much so that their identity is often completely wrapped up in it. Their work values might differ significantly from the work values of their ancestors, who took the Protestant work ethic seriously, but they have not replaced work values with leisure values. Instead, it seems that work values have permeated their leisure values. Increasingly, people's lives today alternate between frenzied work and frenzied play. Rest has been driven out of leisure.

Those who identify with their work and those who distance themselves from it, those who are overworked (either out of need to survive, or out of need to succeed) and those who might be underworked—all need to reflect carefully on the right relationship between work and leisure. Aristotle was right when he maintained that a person needs to learn, "not only to work well, but to use leisure well."[54] But does it make sense to speak of "using leisure well"? Should not leisure activities be free and are they not by definition enjoyable because they are primarily done for their own sake? But not all enjoyable activities are desirable. The leisure character of an activity says nothing about its desirability. As I have defined it, alcohol consumption is a leisure activity, but that does not mean that it should be encouraged. On reflection most of us would agree that the point is, not simply to enjoy, but to enjoy well.

But how does one learn to enjoy leisure well? Indeed, how does one come to know what good leisure is? A theory of the good use of leisure must rest on a theory of human beings and their fundamental needs. Anticipating what I will say later on fundamental needs,[55] I will venture to say now that good "leisurers" are those who know how to enjoy nature, the free exercise of their abilities, fellowship with one another, and, above all, who know how to delight in communion with God.

Work and Worship

Enjoyment of the beauty of nature, delight in the exercise and development of one's own skills, and appreciation of fellowship with one another (especially relations between the sexes) are three fundamental aspects of leisure activity. These activities play an indispensable role in the lives of human beings, both directly and through the influence they exert on the work experience. But space does not permit me to investigate further their nature and their function.[56] I will only address the relationship between work and the central leisure activity in a Christian concept of the good life: communion with God.

Communion with God

The key to our humanness, as I said above, is God's relation to us. The key to our true identity is our communion with God through the Spirit. The question can be posed, do we need to commune with God in the special activity of worship? Since work is a fundamental dimension of human existence, why not simply commune with God by obediently doing God's will? Should not worship be simply a way of life in the world?[57] Why take the time to go to a "secret place" (Matt. 6:6)?

No doubt, Christians should worship God with their whole existence, work included (see Rom. 12:1f.). As they serve God through their work, they worship God.[58] But God did not create human beings simply to be servants but above all to be God's children and friends. As much as they need to do God's will, so also they need to enjoy God's presence. In order to be truly who they are, they need periodic moments of time in which God's commands and their tasks will disappear from the forefront of their consciousness and in which God will be there for them and they will be there for God—to adore the God of loving holiness and to thank and pray to the God of holy love.

The need for a special time of communion with God is grounded in the Christian experience of salvation. At the center of the Christian life lies, not a change in the individual's self-perception, not liberation from sin and impartation of a new ethical orientation in life, not even an anticipatory transformation into a new creation—at the center lies something much deeper, which underlies all these aspects of the Christian life. It is the union of human beings with the Son of God through the Spirit (see 1 Cor. 1:9). Paul describes this union with metaphors that underscore its most intimate nature, such as the metaphor of "being made to drink" of the Spirit (1 Cor. 12:13). When Christians commune with God in worship, they come to drink from that fountain their very life as Christians and hence their identity as human beings depends on. At the same time, in worship they anticipate the enjoyment of God in the new creation where they will communally dwell in the triune God and the triune God will dwell in them (see Rev. 21:22; John 17:21).[59]

The character of charisms reflects the necessity of worship as a separate activity. In the New Testament we encounter, not only gifts that correspond to action in the world (such as the gifts of evangelism and healing), but also gifts that correspond to individual or communal communion with God (such as the gift of singing hymns and speaking in tongues). The Spirit inspires and gifts people not only to work but also to enjoy "festive companionship" with God.[60]

Alternating Activities

Much of Christian tradition over the centuries subordinated work to leisure. Work had no other value than as a means to make people's communion with God possible by keeping their bodies alive and their souls pure.[61] Such an understanding of the relationship between work and leisure is a Christian reflection of Aristotle's influential thesis: "We are busy that we may have leisure."[62] Medieval theologians followed it by making the *vita activa* completely subservient to the *vita contemplativa*.[63] As for Aristotle the purpose of mundane work was the contemplation of truth, so for medieval theologians the only real purpose of work was the contemplation of God.

In modern times the medieval relationship between the *vita activa* and the *vita contemplativa* has been inverted, both in theory and in practice.[64] Practical action, not contemplation, has become the way to truth. Similarly, work has gained the upper hand over leisure. Partly under the influence of Protestantism, which made worship subservient to active life in the world, there arose in Western societies the persistent belief both "that all leisure must be earned by work and good works" and that, while leisure is enjoyed, "it must be seen in the context of future work and good works."[65] Seen from the perspective of a world dominated by work, leisure—and above all, its central aspect, communion with God—was something "wholly fortuitous and strange, without rhyme or reason."[66]

In opposition to both Greek and traditional Christian depreciation of work and modern disparagement of leisure, a Christian theology faithful to its biblical sources will refuse either to subordinate work completely to leisure or to make leisure fully subservient to work. Instead, one must conceive of work and leisure as alternating activities. The anthropological foundation of the "alternation model" lies in the double fact that relation to God is the key to our humanness (and hence our relation to God, the key to our identity) and that work is a fundamental dimension of our existence.[67] The pneumatological foundation is the fact that charisms include both liturgical activities and activities in the world, and that there seems to be no hierarchical ordering of charisms that corresponds to "spiritual" tasks, on the one hand, and to "secular" tasks on the other (see 1 Cor. 12:28 and Rom. 12:7).[68] If God's Spirit inspires people both to work and to worship without systematically preferring either one or the other, then both work and worship must be fundamental activities of human beings that cannot be subordinated to each other. N. Wolterstorff has rightly maintained that the "rhythmic alternation between work and worship, labor and liturgy is one of the significant distinguishing features of a Christian's way of being-in-the-world."[69]

If work and leisure follow in rhythmic alternation, then they limit each other. Since the Christian life consists of both work and leisure, it cannot consist of either of the two exclusively. Augustine rightly observed that "no one ought to be so leisured as to take no thought of his neighbor, nor so active as to feel no need for the contemplation of God."[70] Those who carelessly enjoy a life of leisure—today still a small minority—need to be reminded, not only that they should not eat, let alone live in luxury, if they do not work (2 Thess. 3:10), but also that they are responsible for caring for the needs of the destitute and, if necessary, working in order to be able to do so (Eph. 4:28; Acts. 20:35). Modern workaholics whose lifestyle relegates leisure to an insignificant activity, on the other hand, need to be aware that by wanting to "gain the whole world" they might "forfeit their life" (Mark 8:36). Many an incessant worker, of course, does not want to gain the whole world at all. He only wants to stay alive in a modern, hypercompetitive economic world. The appeals to limit work in these cases will understandably remain ineffective in spite of people's good will. Hence the need for systemic changes to support individuals' decision to follow the moral appeals.

Interpendent Activities

It is not enough, however, to say that both work and leisure need to be pursued as valuable activities in their own right. Because work and leisure are not only alternating, but also *interdependent* activities we also need to ask what positive relationship they have to each other and how they should influence each other.

Although work can be done for its own sake, by definition the end of work lies outside work itself. One important purpose of work is to make leisure possible. In economically developed countries it is taken for granted that one need not work more than forty hours a week. If people are overworked, it is for the most part because they have a wrong attitude either to work or to their desires, not because they cannot otherwise meet their basic needs. For many people in economically developing countries, however, life is "all work, and no play." In that the Sabbath commandment for the first time in history interrupted work with regular periods of rest, it liberated human beings from enslavement to work. This liberating aspect of the Sabbath commandment should be preserved in a Christian theology of work through the insistence that leisure be an inalienable right of every person.[71] The right to leisure is a must, for their survival as workers, and for the protection of their dignity as persons created for communion with God. It presupposes that people can meet their basic needs without having to forego leisure. The right to leisure implies, therefore, the corresponding right to sustenance for all those who are willing to work "six days a week" (see Exod. 20:9).

As a form of leisure activity, worship is clearly an end in itself. But it is by no means a world apart. It influences the work experience. This holds true even if the "influence is more likely to be *from* work experience and attitudes *to* leisure experience and attitudes than the other way round, mainly because the work sphere is both more structured and more basic to life itself."[72] Some recent sociological studies in Australia seem to establish a significant correlation between the strength of individuals' religious orientation and their attitudes toward work.[73]

Communion with God in worship establishes the context of meaning that gives work ultimate significance. One dimension of worship is an anticipatory celebration of the eschatological *shalom*, in which human beings will live in peace with themselves and in fellowship with nature, one another, and God. As a time for "participation in the Spirit that unites what is below and what is above,"[74] worship also inspires people to do their work as creative activity in the service of God's new creation. And finally, the presence of God in worship transforms people so that they can advance through work "that transfiguration of the whole universe which is the coming of the Kingdom of God."[75]

It would be inadequate, however, to conceive of communion with God's Spirit in worship merely as a background experience that influences work from the outside, so to speak. For worship only deepens the continuous presence of the Spirit in the life of a

Christian. To live the Christian life means for Christ to live his life in a person through the Spirit (see Gal. 2:20; 5:22ff). A Christian does not work out of an experience of the Spirit that belongs to the past (a past Sunday experience). She works through the power of the Spirit that is now active in her. The presence of God's Spirit permeates her work experience. She works *in the Spirit*.

SPIRIT, WORK, AND ENVIRONMENT
Work and Nature

In emerging postindustrial information societies, human work seems increasingly dissociated from nature. In both agricultural and industrial societies, for the majority of the population, work was directly related to nature (either in its "natural" or a fabricated form). Speaking from the context of the industrial society, Marx could say without much exaggeration that "nature provides labour with [the] *means of life*" since "labour cannot live without objects on which to operate."[76] In information societies this situation has radically changed. The strategic resource in the world of work is no longer "natural" or fabricated nature, but knowledge. As emerging industrial societies worked themselves out of the agricultural business and into the manufacturing business, so emerging information societies are working themselves "out of the manufacturing business and into the thinking business."[77] Work is increasingly removed from nature.

Given the diminishing importance of nature as an object of work, one might be tempted to think that the discussion of the relationship between work and nature belongs to a bygone era. For several reasons, however, this is not the case. First, information societies are emerging only from highly developed industrial societies, and even in those societies they are still partly unrealized. The majority of the world population today lives in predominantly agricultural and industrial societies. For them, primary or secondary raw materials are still the main resource they work with. Second, even in information societies, agricultural and manufacturing work are an indispensable precondition for every other kind of work. As the expanding contemporary ecological crisis indicates, societies in which the "thinking business" is becoming the primary form of work are not intervening less in nature, but are only doing so more efficiently.

Human work is and always will remain essentially related to nature, because human beings are essentially "natural beings" who can live only in constant interchange with their natural environment. This fact is a sufficient justification for discussing the relationship between work and nature in a theology of work. The contemporary ecological crisis (which I described briefly in Chapter 2[78]) makes such treatment, moreover, imperative.

The charge leveled against biblical tradition—that it contributed significantly to the present ecological crisis—is an additional reason why no theology of work can avoid discussing the relation of work to nature, at least not a theology of work that wants to bear in mind its own practical consequences. Ever since the sixteenth century, Genesis 1:26-28 has served as a *locus classicus* for a Christian understanding of work in general and for the relationship between work and nature in particular.[79] The text states that human beings alone were created in the image of God and that the purpose of their work is to rule over the rest of the creation, to "subdue" the earth and "have dominion" over animal life. Both the uniqueness of human beings within the creation and the subduing of the earth as the purpose of their work are blamed as significant causes of the contemporary destruction of natural environment.

No doubt, the idea of *dominium terrae* played a significant role in *justifying* modern striving to conquer nature, with all its calamitous consequences. At the beginning of modern times, influential philosophers used this notion to express their belief that the rule over nature through technology was the most important goal of human beings.[80] Theologians in the Western tradition often went along, and saw in Genesis 1 the mandate for unfettered technological development. To give a recent example, in *Laborem Exercens*, John Paul II claims that, as in the "periods of 'acceleration' in economic life" human beings became more and more masters of the earth through "the progress of science and technology," they "remain *in every case and at every phase of this process* within the Creator's original ordering" expressed in the first creation account.[81]

As I will show later, to use Genesis 1 to sanction technological development *tout court* is to misuse this passage. But first I need to discuss the broader question of the proper relationship of work and nature. The answer to this question depends, in turn, on the character of the relation between human beings and nature. So I begin by discussing the question of the "naturalness" of human beings.

Spirit, Persons, and Nature

The anthropological foundation for the modern ecologically destructive relation of human beings to nature is the sharp dichotomy between the soul on the one hand, and the body and nonhuman environment on the other.

Descartes, who thought human beings should render themselves "lords and possessors of nature" in order to "enjoy without any trouble the fruits of the earth, and all its comforts,"[82] provides a representative and influential example of such a dichotomy. The human "I" is for Descartes "a thinking and unextended thing." It is identical with the mind and constitutes a person. The "body," on the other hand, is "an extended and unthinking thing."[83] Despite the fact that Descartes grants the mutual influence of the mind and the body, a person cannot be said to be a body as she is a mind, but only to possess a body. The "I" stands against the body and all other extended objects with the task of ruling over them and possessing them.

Following philosophers like Descartes (and Plato before him), Christian theologians have for centuries stripped the human spirit of everything corporeal and emptied corporeality of everything spiritual. In spite of its long tradition, such an understanding of human beings and of nonhuman nature is not biblical. As careful study of Genesis 1 shows, while created in the image of God and called to have dominion over creation, human beings are not set against the rest of creation but are embedded in it (see also Ps. 104).[84] They do, of course, have a special position in creation, for they alone are created in the image of God. But this special position in nature does not imply a denial of their basic naturalness.[85] We have to maintain not merely that human beings have bodies, but that they are bodies (though, of course, not reducing them to bodies). The body is not simply an extrinsic instrument a human being uses; it is an integral part of their identity as a particular human being.

What unites human beings and the nonhuman creation is, however, not only their common material creatureliness, but even more significantly, the presence of God's Spirit in both of them.[86] During the Reformation, Calvin in particular stressed the presence of the divine Spirit in nature: "For it is the Spirit who, everywhere diffused, sustains all things, causes them to grow, and quickens them in heaven and in earth." Calvin continues by saying that the Spirit transfuses "into all things his energy" and breathes "into them essence, life, and movement."[87]

Calvin was drawing on aspects of biblical tradition that have been neglected until recently. We read in Job 34:14f. (cf. Ps. 104:29-30), for instance, that the presence of the Spirit of God gives life both to human beings and to all other living creatures:

> If he [God] should take back his spirit to himself,
>   and gather to himself his breath,
> all flesh would perish together,
>   and man would return to dust.

Although it does not say so explicitly, the New Testament also implies that the Spirit is present in creation. Paul says that the creation is "groaning in travail" because it is subject to transitoriness, and that it "waits with eager longing" to be "set free from its bondage to decay and obtain the glorious liberty of the children of God" (Rom. 8:19-22). Even as metaphorical language, talk about the "groaning" and "eager longing" of the creation for liberation is unintelligible if the creation is perceived merely as an extended object. These metaphors suppose that creation has some "feeling" of a gap between its present existence and its future destination. In the context, this "feeling" is best made intelligible with reference to God's Spirit (who is later described as groaning in human beings [v. 26]).

Work as Cooperation with Nature

The presence of the Spirit both in human beings and in nature, which establishes their fundamental unity, has two important consequences for the relationship between work and nature.

Creation as an End in Itself

If the Spirit is present in nature, then nature must be considered partly an end in itself. It has value independent of its service to human beings. This is implied also in the fact that God has a distinct soteriological relation to nature (which is, however, not separable from God's soteriological relation to human beings). Paul makes that clear in Romans 8: as creation is suffering the consequences of human sin, so will it also participate in the "liberty of the children of God" (v. 21). Nature is also heir to the eschatological glory. Similarly, some Old Testament passages speak about a special relationship of God to all living beings. According to Genesis, after the deluge, God established a covenant not only with Noah and his descendants but also "with every living creature that is with you . . . as many as came out of the ark" (Gen. 9:9f.). Animals belong to the same covenant as human beings; they are perceived as "partners of God in their own right."[88]

If nature has independent value, then it can be treated neither as merely a resource from which material wealth is created (as Smith implied), nor as something that can achieve its goal only if humanized by work (as Marx maintained).[89] Human beings are responsible for respecting nature in its specific creatureliness. Respect for nature does not necessarily imply that they should cease using it as a means to certain ends, for nature has no inherent sanctity (the language of the "holy" should be kept for the sphere of religion where it properly belongs[90]). Without an instrumental relation to the nonhuman creation, human beings could not survive, let alone live dignified lives. Respect for the nonhuman creation does not require a purely aesthetic relation to nature,[91] but it does imply that all human work has to include an element of caring for creation.

According to Genesis 2, human beings were given the tasks of "working" and "taking care" of the Garden (v. 15). These are not two separate kinds of activities, but rather two aspects of all human work.[92] Interpreting Genesis 2:15 in an agricultural setting, Luther said: "These two things must be done together; that is, the land is not only tilled, but what has been tilled is also guarded."[93] All work must have not only a productive but also a protective aspect. Economic systems must therefore be integrated into the given biological systems of ecological interdependence.

Caring as Perfecting

Protection from ecological imbalance and irreversible damage is an important dimension of care for the nonhuman creation. Taken by itself, however, this dimension of care is inadequate because it is too static and oriented exclusively to the past (restoration) and present (protection). An adequate concept of care for nature must have a dynamic, future-oriented dimension. It must take into account that there is a *history of nature* that necessarily accompanies the human encounter with nature. The question is not simply how to preserve the environment in the state in which it presently exists, but *how to preserve the naturalness of nonhuman nature that is subject to change because of the human relation to it.*

If the history of the nonhuman creation that results from human work is not to degenerate into a history of the denaturalization of nature, then human beings will have to pay heed to the hidden tendencies of nature in order to create the conditions that will *facilitate* the realization of nature's potential. Hence the need for human beings to cooperate with the nonhuman creation as they work on it. We should not think of the worker only as imposing order on recalcitrant nature, but also as engaging "in a sort of dialogue with her material" by which "she lovingly coaxes it into revealing its potential."[94] The point of this dialogue is to help nature in a small and broken way to grow into ever greater correspondence to the state into which it will be transformed. In this way we will be able to pass on a better earth to future generations.[95]

A dynamic, cooperative relation to the natural environment is implied in a pneumatological understanding of work. God's Spirit is present in the nonhuman creation that is the object of work, and prompts its longing for liberation. The same Spirit gives inspiration and guidance to working people. The particular experience of the Spirit human beings have (the Spirit as the firstfruits of salvation) serves not to separate them from the nonhuman environment but to unite them with it. For the possession of the Spirit as the firstfruits of salvation leads human beings into solidarity with the nonhuman creation.[96] As they work under the inspiration of the Spirit, they cooperate with it, mindful of its longing to participate in the "glorious liberty of the children of God" (Rom. 8:21).

On Subduing the Earth

How does the emphasis on caring for nature square with God's command to subdue the earth (Gen. 1:26–28)? The text itself makes it clear that in the exercise of dominion over nature, human beings are given a twofold responsibility. They are responsible to God who created them and placed them within the fellowship of creation and they are responsible to their fellow human beings.

First, human beings received the task to subdue the earth as creatures made in the *image of God*. In the exercise of *dominium terrae* they function as God's stewards who are responsible "to rule over the earth in a way that God rules over his world."[97] As Psalm 104 shows, God does not violently ravish nature, but sustains it with providential care. Human beings may not "corrupt by abuse" the nonhuman creation.[98] The purpose of human dominion over nature is the preservation of the integrity of the nonhuman creation, not simply the satisfaction of human needs and wants.

To assert that caring is an essential characteristic of *dominium terrae* might seem to conflict with the meaning of the Hebrew words for dominion used in the Genesis passage (*rdh* and *kbš*). Both verbs have "an explicitly harsh, almost bellicose tone."[99] But the question is whether they retain the same aggressive tone in Genesis 1. First, one should note that ruling (*rdh*) over animals excludes the option of killing them for food.[100] Together with human beings, animals are given vegetation for food (Gen. 1:29). Furthermore, the use of the verb *kbš* elsewhere in the related texts (Num. 32:20-32; Josh. 18:1f.; 1 Chron. 22:18f.) suggests that its meaning in Genesis might not be the commoner "to trample under foot," but "to use the earth for stock-farming and settlement."[101] Finally, it must be kept in mind that through the command to subdue the earth the *blessing* of God will come to realization, and therefore that the command cannot be taken to sanction the destruction of creation (see Gen. 1:28).[102] A careful exegesis of Genesis 1:26ff. reveals that human caring, which reflects God's providential care, is an integral part of *dominium terrae*.

Second, human beings were created in the image of God as a community of men and women, not as isolated individuals (Gen. 1:26ff.). As God, in whose image they were created, is not a monad but a fellowship of three divine persons, so also human beings are not solitary but social beings. Human beings, therefore, have to exercise *dominium terrae* in *responsibility to the whole human community*, the global community in the sequence of generations.[103]

The rootless self-gratification of individuals, single nations, and whole global generations with respect to nature conflicts with the communal dimension of *dominium terræ*.

The twofold human responsibility—to God and to the universal human community—indicates that the exercise of dominion over the earth in the biblical tradition is not simply a question of technological power, as philosophers at the beginning of modernity would have us believe. Bacon recognized that "man fell . . . from his state of innocency and from his dominion over creation" as a result of falling into sin. But whereas he believed that innocence can be regained "by religion and faith," he maintained that for regaining dominion, the "arts and sciences" would suffice.[104] He granted that the exercise of technological power should be governed "by sound reason and true religion."[105] Yet it did not take long before modern societies emancipated themselves from religion, and dominion became perverted into an exercise of brute technological power over recalcitrant nature.

We need to rediscover the religious and moral dimensions of dominion over nature. Luther rightly stressed that the dominion stemming merely from human industry and skill is an inferior dominion. Only "upright" human beings can exercise true dominion over nature.[106] This is the point of the statement that Jesus "was with the wild beasts" during his temptation, yet did not need to tame them (Mark 1:13): his communion with the wild beasts—the communion of him who conquered the temptation to misuse power for self-aggrandizement—anticipated the eschatological peace between human beings and creation that will be the fruit of righteousness (see Isa. 11:6-8; 65:25).

SPIRIT, WORK, AND HUMAN NEEDS

Mere observation suffices to show that as a rule people work in response to certain perceived needs, primarily needs for material products. They work to acquire the things necessary for sheer survival. Most people throughout history worked day in and day out, the whole of their lives, in order to satisfy their most basic needs. And so do many people today. But even those more fortunate who could reduce their workday to a fraction of its duration and still be able to meet their basic needs in fact work the whole time for the most part in order to satisfy their material needs. They work for products they *feel* they need in order to live with dignity. For most people, needs are the end, work a necessary means to achieve it.

It is a tribute to the realism of the Bible that it does not simply religiously transfigure work into divine service (as in the Mesopotamian myths of biblical times[107]) or into a means of human self-development (as does Marx), but relates it very soberly to mundane human needs. No doubt, we can and should understand work as participation in the divine activity of transforming the world, and we must say that through work human beings perfect (or degrade) their capacities. Yet, as we think of those higher purposes of human work, we cannot "gloss over" that the first thing at issue in all fields of human work is humans' need "to earn their daily bread and a little more."[108] In the Bible and in the first centuries of Christian tradition, meeting one's needs and the needs of one's community (especially its underprivileged members) was clearly the most important purpose of work.[109]

Expanding Needs

If human beings worked simply in order to satisfy their needs, little theological reflection on human needs would be required. The biblical injunction to work for our own and others' needs would suffice (see 1 Thess. 4:12b [NASB]; Eph. 4:28). For the problem at hand would be to make sure that no able people live unjustifiably from other people's work, and that those who are unable to work still have their basic needs met.

But the relationship between work and human needs is more complex. We can see that as soon as we ask, Why do people who work in highly productive modern economies spend as much (and often more!) time securing their "product-needs" as people who live in economically very underdeveloped countries?[110] Part of the reason at least is that their *felt* product-needs are greater. As the wealth of societies grows, the needs of their citizens develop. From the "three humble necessities" (food, clothing, and lodging), they swell, as Adam Smith pointed out, to include various "conveniences according to the nicety and delicacy of taste."[111]

The *dynamic character of needs* is a specifically human phenomenon.[112] It is grounded in the permanent self-transcendence of human beings. What human beings need is always beyond the boundary of what they actually have and are. So they live in an endless spiral in which today's desires glide into tomorrow's needs. Depending on the culture, the movement of the spiral might be slower or faster. The upward-moving spiral itself seems inherent to the human condition.

As we reflect theologically on human work we must not only notice that human needs are inherently dynamic, expanding. We must also grasp that at the root of modern economic life lies a stress on the *permanent expansion of human needs*. True, because of the dynamic character of specifically human needs, the needs of cultures before the appearance of capitalism and socialism were not permanently in equilibrium, either.[113] But in most of these cultures a person of virtue was a person of few needs. Weighty moral arguments were marshaled to counter the multiplication of product-needs. With the appearance of capitalism the situation changed. The liberal utopia—and the same became true of the later socialist utopia—is premised on the permanent expansion of human needs. Affirmation of the incesssant proliferation of product-needs is integral to the consciousness of modern peoples: insatiability seems inseparable from modernity.[114] This is partly so because the modernity itself rests on a type of economy that creates human needs in order to satisfy its own "need" for permanent growth.[115]

Yet insatiability is being challenged by the very success of modern economies in providing people with consumer goods. The ecological crisis casts a dark shadow on the unlimited proliferation of product-needs. It reminds us powerfully of the lessons about the dangers of wealth from the Christian spiritual tradition, which Smith and Marx (and with them the whole civilizations of the so-called First and Second Worlds) thought they could disregard. The ecological crisis forces us to seriously consider limiting the expansion of product-needs.

To suggest that one might need to limit the expansion of human product-needs does not imply an affirmation of zero-growth economy. There are significant ways economies can grow without endangering the natural environment. But it does mean that economic growth must be attuned to ecological health.

Fundamental Needs

But how should one determine whether product-needs should be curtailed or not? And if they should, how should this be done? I suggest that we need to place product-needs into the larger framework of a theory of human needs.

Before I briefly discuss fundamental human needs and indicate how they relate to product-needs, I should point out one formal feature of a Christian theory of human needs. Because it presupposes a normative human nature, in a Christian theory of needs, the distinction between true and false needs will be not only legitimate but necessary. It will, therefore, be at odds with the reaction of the youth in Plato's *Republic* who, at the suggestion that "some pleasures are the satisfaction of good and noble desires, and other, evil desires," "shakes his head and says that they are all alike, and that one is as good as another."[116] It will also be at odds with many modern social philosophers who consider the distinction between true and false needs merely a tool—and a dangerous one at that—in ideological debates.[117]

Ways of Limiting Product-Needs

There are many ways one could attempt to limit the expansion of product-needs (if such limitation is necessary). The way dominant throughout church history has been by preaching one or another form of *asceticism*. But asceticism denies the dynamic nature of specifically human needs and misinterprets the biblical tradition as well. By affirming the goodness of creation, biblical tradition encourages enjoyment of it (see Matt. 11:19; 1 Tim. 6:17).

The other way to limit product-needs would be for a government to impose on its citizens what it considered to be an acceptable structure of needs. Commenting on this idea, Ignatieff rejects it by maintaining that "there are few presumptions in human relations more dangerous than the idea that one knows what another human being needs better than they do themselves."[118] But surely the problem cannot lie in *knowing*. What can be wrong in a specialist's knowing better what I need in order to reach a set goal (like health or retirement security) than I know myself? There is nothing wrong in their knowing, but there would be everything wrong in their having the *power* to implement their knowledge regardless of my preferences. The fearful thing about a government imposing a structure of needs on its citizens is its resulting dictatorship over needs, which is one of the worst forms of dictatorship imaginable.

If one wants to respect the dynamic character of human needs and freedom in need-satisfaction, then it seems most sensible to attempt to limit product-needs by helping people discover and develop other *fundamental needs* (to be described later) that do not require satisfaction through human material production. The possibility of satisfying those fundamental needs would at the same time function as the criterion for determining to what degree one can responsibly indulge product-needs. If striving for the satisfaction of product-needs makes it impossible or difficult to satisfy non-product-needs that are fundamental to people's humanity, then the limitation of product-needs is in order. If such is not the case, however, then product-needs can be validly expanded.

Fundamental Needs

Fundamental non-product-needs are objectively rooted in the nature of human beings as creatures made in the image of God, but are subjectively awakened and satisfied by the activity of the Spirit of God. The same Spirit that empowers people to work in order to satisfy their product-needs also motivates them to satisfy their fundamental needs that might limit their product-needs. A pneumatological understanding of work will therefore have to call into question both working for product-needs at the expense of satisfying non-product-needs (consumerism), and satisfying non-product-needs at the expense of working for product-needs (asceticism).

What are the fundamental human needs that should function as the criteria for legitimate expansion of product-needs? I will briefly suggest four fundamental, anthropologically grounded needs and point out their relation to the Spirit of God.[119]

(1) The most fundamental of all human needs is the need for communion with God. "Although my real need was for you, my God, who are the food of the soul," writes Augustine, "I was not aware of this hunger. I felt no need for the food that does not perish, not because I had my fill of it, but because the more I was starved for it, the less palatable it seemed. Because of this, my soul felt sick."[120] Mary, who sat at the Lord's feet and listened to his teaching, had chosen "the one thing needful" (Luke 10:42), the one

thing that gives health to the "soul." The experience of the fullness of life in God's presence guards people from an insatiable desire for material wealth (see Isa. 55:2). On the other hand, when the presence of the Spirit awakens and satisfies the need for God, it "creates a sphere of real wealth and superfluity, even in the midst of the direst material want" and sets up an area "where usefulness is forgotten and generosity reigns."[121]

(2) The growth of product-needs that endanger the natural environment can be limited if human beings develop the need for *solidarity with nature*. The presence of the Spirit can play an important role in the development and nurturing of this need. As mentioned earlier, the Spirit is the power of eschatological salvation in which both human beings and nature will participate.[122] As much as the experience of the Spirit separates human beings from nature (for they alone have the Spirit as the firstfruits of the future glory), it also creates in them a feeling of solidarity with nature (for it makes them conscious of their destiny to belong with nature to the one eschatological new creation). For this reason, those who have experienced the power of the Holy Spirit cannot be indifferent to the destruction of the nonhuman creation.

(3) Human beings should also develop a need for tending to the well-being of one another. Those who respond to the Spirit's call to come from the darkness into God's marvelous light are incorporated by the Spirit into *one body in Christ* (1 Cor. 12:13) and bound with the bond of love. It is well known that two people who love each other "will react to each other's needs and their own indiscriminately. Indeed the other's need *is* his own need."[123] When the spirit of fellowship is present in the body of Christ, a person is liberated from an individualistic search for satisfaction of product-needs because the needs of each individual become the needs of the community and the needs of the community, the needs of each individual.[124]

(4) The need for human development functions as a further criterion for the proliferation of product-needs. In the New Testament the most significant development occurs at two levels: it consists of the development of moral capacities and of practical and intellectual skills. At both of these levels, human development occurs through the operation of the Spirit—at the first level, as nurture of the "fruit of the Spirit," and at the second, as the

kindling afresh of the "gifts of the Spirit."[125] The expansion of product-needs that goes beyond what is necessary for healthy subsistence is valid only if it does not happen at the expense of the worker's opportunity for development.

(5) All four fundamental needs are grounded on the single, underlying universal need, the need for the *new creation, which is the kingdom of freedom*. In relation to the *need for God*, the kingdom of freedom is a kingdom of perfect fellowship with God, of seeing "face to face" and of understanding as fully as one is "fully understood" (1 Cor. 13:12). In relation to the need for *solidarity with nature*, it is the kingdom of peace between human beings and nature liberated from corruptibility and the kingdom in which human beings jointly participate with nature in God's glory (Rom. 8:19ff.; Isa. 11:6f.; 65:25). In relation to the *need for fellow human beings*, it is the kingdom of unadulterated fellowship with one another, of pure "love, which binds everything together in perfect harmony" (Col. 3:14). In relation to the *need for personal development*, it is the kingdom in which life is "realized only to be opened up to as yet unrealized possibilities."[126]

The need for the new creation is the broadest context in which the expansion and limitation of product-needs should be placed. The striving for satisfaction of product-needs is legitimate only if it does not hinder satisfaction of this fundamental need. But new creation is more than just an ethical criterion for the satisfaction of product-needs. The need for the new creation, awakened and kept alive by the Spirit of God, and the anticipatory experience of new creation in the presence of the Spirit, protects us from being caught in the endless spiral of satisfaction of product-needs. For Jesus said, "Seek first his kingdom and his righteousness, and all these things shall be yours as well" (Matt. 6:33).

EXCURSUS: SPIRIT, WORK, AND UNEMPLOYMENT

Economically developed countries today do not have to live with the high unemployment rates of the early eighties. It would be a mistake, however, to bypass the problem of unemployment. For even if we might disagree about whether rapid technological development will lead to high unemployment rates without the appropriate structural, social, and economic changes,[127] we can expect with a high degree of certainty that another wave of increased unemployment rates will strike.

For economically developed countries, high unemployment rates are a future threat; for many economically developing countries, however, they are a present reality. In the relatively small country of Yugoslavia (which has 23 million inhabitants), for instance, more than 1.5 million people are looking for jobs. If the government implements badly needed structural reforms of the economy, it is expected that another 1.5 million will be out of a job. Given the high personal and social costs of unemployment,[128] there is great reason for concern.

As I have done so far in this book, here also I will leave aside questions of economic and social policy (not, of course, because I consider them unimportant, but because I consider myself incompetent to address them). I will make only two comments about resources a pneumatological understanding of work has for dealing with unemployment. These will be the only explicit comments I will make about the problem of unemployment. Compared with the gravity of the problem, this might seem too little. But if it is true that the problem of unemployment is inseparable from the problem of good work,[129] then my theology of work as a whole is an indirect contribution to solving the problem of unemployment. I leave to the reader the task of drawing the conclusions about the problem of unemployment from what I have to say about the relation of work to human beings—to their nature, needs, and leisure—and to the natural environment, and from what I will have to say about the alienation and humanization of work.

Since a pneumatological understanding of work supposes a synchronic plurality of charismatically based activities, it makes an important contribution by helping people deal subjectively with the problem of unemployment. For many of the unemployed, the financial difficulties caused by unemployment are no bigger problems than "the moral grievance of vacuity and boredom" and "the spiritual grievance of being allowed no opportunity of contributing to the general life and welfare of the community."[130] The historic Oxford Conference (1937) stresses rightly that prolonged unemployment "tends to create in the mind of the unemployed person a sense of uselessness or even of being a nuisance, and to empty his life of any meaning." It continues by noting that "this situation cannot be met by measures of unemployment assistance, because it is the lack of significant activity which tends to destroy his human self-respect."[131]

As developed by later Lutheranism, the vocational understanding of work seriously aggravates these problems, since it reduces significant activity of human beings to employment.[132] A pneumatological understanding of work is better equipped to deal with the situation. Since an unemployed person has not been deprived of all charisms, in a pneumatological understanding of work he is not left without a divinely appointed, significant activity. Rather he is merely unable to serve God and his fellow creatures with one particular—possibly even not the most important—one of her various charisms. To be unemployed need not mean being without work, but can mean being free for other significant kinds of work.

The problem of unemployment is certainly not yet solved when we help the unemployed deal with it subjectively. We need to deal with it also objectively, either by creating new jobs or by distributing available work more evenly. The pneumatological understanding of work contributes indirectly to a more even distribution of work by calling into question the excessive value put on employment. First, it relativizes employment by making it only one of the many likewise valuable ways in which a person can be active in the secular realm. Second, it relativizes employment by placing it on the same footing as the ecclesiastical activities of individuals. Employment need not be necessarily *the* task of an ordinary Christian (as ministry is *the* task of a professional minister), as the vocational understanding of work maintains. Employment can be but *a* task set by God. As I have defined it, *charism* relates to activities both in the church and in the world. True, Luther applied the notion of vocation both to secular and to ecclesiastical activities. But because he did not operate with a synchronic plurality of vocations, he could only imagine an *inter*personal distribution of vocations in the world and in the church: one individual is *either* called to ministry *or* to secular work. The pneumatological understanding of work operates with an *intra*personal distribution of activities. Without denying the need for specialization, it stresses that every Christian can simultaneously have different charisms and contribute to the edification of the church and the transformation of the world.

CHAPTER 6

Alienation and Humanization of Work

With the phrase "alienating work" I am referring to a significant discrepancy between what work should be as a fundamental dimension of human existence and how it is actually performed and experienced by workers. This discrepancy came to be felt most acutely at the beginning of the nineteenth century, due to changes in anthropology and the character of human work. Work was pronounced to be the very essence of human beings (Marx), and yet for sixteen or more hours a day it stupefied people's minds and ravaged their bodies. Since that time, economically developed nations have done much to alleviate the workers' plight. Various forms of humanization of work and labor-protecting laws have been implemented to make the nature of work better match the nature of workers. Yet, as the unabated dissatisfaction with work worldwide indicates, alienation, like a dark shadow, still persistently accompanies most of human work.

In this chapter I intend to draw the implications of the pneumatological understanding of work for the seamy underside of the modern world of work.[1] I will examine the principal forms of alienating work in the modern world and place them against the backdrop of humanized work as the goal toward which a Christian should strive. But before approaching this central issue, I will attempt to bring some semblance of order to the clutter that surrounds it. So, I will first examine whether alienation refers exclusively to the subjective experiences of workers or to an objective reality. This is the central and most disputed question in the modern debate about the character of alienation in work. Second, I will examine the reasons why a Christian should (or should not) be concerned with the problem of alienating work.

CHARACTER OF ALIENATION

The question of whether or not alienation is an objective reality that can be established without reference to the affective response of the worker to her work either looms large in the modern debate about alienating work or leads a vital underground existence if authors choose not to discuss it explicitly. The reason is not hard to find. The answer to this question will determine one's basic approach to the problem of alienating work: whether one will strive only to keep workers satisfied or endeavor to help them work in accordance with their nature (which will, one hopes, give them good reason to be satisfied).

Job Satisfaction

There is a strong tendency today to treat alienation in work not as an objective discrepancy between the nature of the worker and the character of his work, but as an overall negative affective response to a particular work role.[2] Hence the problem of alienation in work becomes the problem of job satisfaction.[3] Referring to one important aspect of alienation in work, Kohn writes: "It is not the fact of being powerless but the *sense* of being powerless that we attempt to measure."[4] Whether it be out of concern for increased productivity, or fear of the spillover of dissatisfaction with work to other spheres of life, or because they consider job satisfaction inherently valuable, modern social scientists concentrate their efforts on identifying the intrinsic and extrinsic factors that influence satisfaction and dissatisfaction with work, and suggesting ways to increase the first and eliminate the second.

One would not want to take issue with this approach, as far as it goes. Since it is indisputable both that "every worker aims at joy in work, just as every human being aims at happiness"[5] and that satisfaction is preferable to frustration with work, one should strive to help a person "find enjoyment in his toil" (Eccles. 6:19). Work is meant to be "pleasant and full of delight."[6] After all, it fills most of the waking hours of the average person. But will the concern with work satisfaction suffice?

One problem with treating alienation in work as a form of dissatisfaction is the observed lack of correlation between the satisfaction of workers and the character of their work. On the one hand, the increase in the substantive complexity of work and the self-directedness of the worker (to name two interrelated aspects of the humanization of work) often do not lead to a corresponding increase in satisfaction for workers.[7] Similarly, one would expect that Japanese workers, who have lifetime employment in community-type organizations where participation and self-development are encouraged, would be the most satisfied. In fact, studies indicate that they might be "the least satisfied workers in the world."[8]

The tendency of people not to be satisfied with their work when they have good reasons to be satisfied does not invalidate treating the problem of alienation as a problem of dissatisfaction. It only makes it evident that the concern with dissatisfaction needs to be extended to other spheres of workers' lives that might be causing their dissatisfaction with work. But there is an opposite tendency, too. People tend to invest their work with deep significance and claim to be satisfied with it even when there are no objective reasons for such attitudes in terms of the actual importance and quality of the work performed. Since people are reluctant to admit failure in life and hence suppress dissatisfaction with their jobs,[9] it is very possible to have a "relation of jolly slavery to work."[10] Alienation in work is not equivalent to dissatisfaction with work;[11] one can obviously be alienated without being dissatisfied.

One will, of course, have to respect peoples' feelings and the job preferences that accompany them, but one will nevertheless be justified in maintaining that these feelings and preferences are objectively wrong. People should not be satisfied because of their alienating jobs, but in spite of them.[12] A theory of alienation that concentrates on work satisfaction cannot make this necessary step beyond people's feelings and preferences.

Nature of Workers and Character of Work

The main reason for the modern preponderance of the subjective conception of alienation is general dislike for the idea that there is a normative human nature (which possesses at least some stable, culturally unconditioned qualities) as is supposed by an objective conception of alienation.[13] But since a Christian theologian has a commitment to a normative anthropology,[14] she will have to insist on the objective character of alienation. Work is alienating when it does not correspond to God's intent for human nature. As I hope to show later, various forms of alienation represent various ways work negates human nature. The primary goal in the struggle against alienating work cannot, therefore, be simply to help people deal with feelings of frustration with work (though this is needed, too). The transformation of the character of work is required, not primarily in order for it to match workers' feelings but for it to correspond to their nature. Without such a transformation of work, alienation in work will be masked, not overcome.

An objective understanding of alienation should, nevertheless, not be formulated without reference to workers' subjective states of consciousness. Alienating work is not just a matter of discrepancy between the character of the work role and the objective nature of the worker. It also concerns workers' attitudes. Take the example of work being treated as a mere means—a form of alienating work I will analyze later.[15] Work can be structured in a way that does not foster treating it as a mere means, but no cooperative work can ever be structured to absolutely exclude this form of alienation. Alienating work sometimes does stem from attitudes of workers that are not caused by the structure of the work they do.

If alienation exists regardless of the objective characteristics of work, that does not mean that alienation is merely a subjective phenomenon. It is subjective in that it is caused by the individual's subjective attitudes. The criteria, however, by which these attitudes are judged as alienating or not do not stem from the individual concerned (as his positive or negative affective response to work would), but are based on certain characteristics of human nature and are therefore objective. To stay with the example given above, to treat work as a mere means—whether one does this because of the objective character of work or one's subjective state of mind—is alienating not simply because it makes work unenjoyable, but because work is a fundamental dimension of human existence.[16]

Often alienation is a matter of discrepancy between work and the individual's *inclinations*, not between work and human nature. With respect to human nature, the tasks of doing theology and administrating a theological institution would, I suspect, be equally non-alienating. Yet a person who has a strong inclination for doing theology but for some noble (or ignoble) reason became an administrator, a job she dreaded, would probably be dissatisfied. One could rightly call such a person self-estranged, though not estranged from herself as a human being (in her nature), but as an individual (in her identity). Reformulating what sociologists of work call the "fit" hypothesis,[17] one can say that reducing alienation in work will always involve both matching work with human nature and matching people of diverse individual inclinations with work roles that suit them (work roles that are, one hopes, matched with human nature).

One can be alienated without noticing it, whether that alienation has a subjective or objective cause. But one cannot feel alienated without being alienated. Although satisfaction with work is not an unmistakable indicator of the absence of alienation, discontent with work is an unmistakable indicator of the presence of alienation. Because self-perception is an integral part of human identity, the person who perceives himself as alienated *is* alienated. *Negative affective response to work* is, therefore, an unmistakable indicator of alienation, even when there is no discord between the character of work and the nature of the worker. The cause of alienation need not, of course, lie in the sphere of life where alienation is felt. Discontent in nonwork areas of life that contribute significantly to the sense of general life satisfaction (such as family relations or sense of purpose in life) often spills over and creates dissatisfaction with work, even if work itself is not objectively alienating.

REASONS FOR CONCERN

In the preceding discussion about the character of alienation in work I simply assumed that a Christian should be concerned about this problem and strive to overcome it. But what are the grounds for concern? In order to answer this question, I will first investigate two reasons commonly given today and then contrast these with reasons implicit in Christian anthropology and soteriology.

Alienating Work, Economic Progress, and Universal Emancipation

One can approach the problem of alienating work from the context of commitment to economic progress, as Adam Smith did. As I have indicated earlier,[18] he considered alienation inseparable from the capitalist economy because of his twofold persuasion about economic proress: first, he considered economic progress a necessary prerequisite for the progress of culture and civilization; and second, he believed that there can be no economic progress without division of labor, which causes alienation. He was persuaded that the good of the laborers is in the long run best served by the alienating division of labor because only the division of labor can create economic progress.

The failure of Taylorism has persuaded Smith's modern followers that at least the minute division of labor will not always result in greater productivity. But because of their basic commitment to economic progress (either as an end in itself or as a necessary means to some other, nobler end) they will as a rule be concerned with the eradication of alienating work only to the extent that the increased satisfaction or skill of workers who do not perform alienating work will contribute to the increase of productivity.

Because of the belief in the goodness of creation, a Christian will basically affirm economic progress (though he will evaluate it in terms of how it relates to the satisfaction of fundamental human needs[19]). If minute division of labor hinders economic progress, this is one good reason to eliminate it. But economic progress cannot be the main reason for Christian concern about alienation of workers. Alienation is problematic, not merely as a potential impediment to increased economic output, but above all as an assault on human nature. The problem with alienating work is not in what it does to production but in what it does to the producer.

In contrast to Smith, Karl Marx was perturbed with what alienating work did to human beings; so much so, that he let a large portion of his theoretical edifice rest on the astonishing but influential claim that alienation in work is the root cause of the entire human predicament: "The whole of human servitude is involved in the relation of the worker to production, and all relations of servitude are but modifications and consequences of this relation."[20] In terms of liberation, this understanding of the human predicament implies that "the emancipation of workers contains universal human emancipation."[21]

Given Marx's anthropological persuasion that work constitutes the "essence" of constantly developing human beings, his reasoning is plausible. As I have argued earlier, in Christian anthropology it is God's personal relation to human beings, not human work (or any other human activity), that constitutes human beings *as* human beings. Consequently, the fundamental form of alienation cannot be alienating work, but alienation from God. Stated in traditional theological language, sin against God has ontological (though not necessarily temporal) priority over all other forms of human sin and misery. The various forms of alienation in human relations to oneself, one's fellow human beings, and nature are ultimately consequences of the fundamental human alienation from God.

The priority of alienation from God over alienation in work has three important implications. First, one should not expect too much from any success one might have in overcoming alienating work. Such success does not reach deep enough into the human predicament to be able to thrust human beings to the heights of universal emancipation. Second, even attempts to humanize work with more modest goals will be less successful than they could be if they concentrate on humanizing work without paying attention to alienation in other spheres of human life; in particular, alienation from God. Since there is a necessary spillover from alienation from God to alienation in work, the full humanization of work requires overcoming alienation from God. For work to be humanized, the working person herself must be "humanized," not least by nurturing her right relation to God. Third, since alienation from God will be overcome only in the new creation, all attempts to humanize work will be crowned with only partial success.

Alienating Work and Anticipation of New Creation

A Christian will strive to reduce alienation in work, because all success in humanizing work anticipates—on a small scale and under the conditions of history—the eschatological new creation.

Since by arguing for understanding work as human proleptic cooperation in God's eschatological transformation of the world,[22] I have implicitly argued for understanding the attempt to reduce alienation along the same lines, I will here only point out some key biblical texts that corroborate these arguments.

Biblical Critique of Alienation

The two central salvific events in the Old and the New Testaments concern at least in part the problem of alienating work. In the Old Testament, God revealed himself to Israel as Yahweh when the Israelites were liberated from *forced labor and slavery* (Exod. 1:13-14). The Israelites relived their liberation every year in ritual form by reciting the "historical credo": "And the Egyptians . . . laid upon us hard labor. . . . Then we cried to the LORD . . . and the LORD brought us out of Egypt . . ." (Deut. 26:6-8). The repeated recital of the history of their liberation was a continual implicit criticism of forced labor among Israelites themselves. The holiness code makes God's liberation of the Israelites the reason why the Israelites should not oppress their neighbors: "And if your brother becomes poor beside you, and sells himself to you, you shall not make him serve as a slave. . . . For they are my servants whom I brought forth out of the land of Egypt. . . . You shall not rule over him with harshness, but shall fear your God" (Lev. 25:39ff.).

The past experience of liberation from forced labor in Egypt led the Old Testament prophets to vehemently attack those who exploit their neighbors. A good example of prophetic criticism of alienated labor is Jeremiah's words to the Davidic king Jehoiakim (Jer. 22:13):

> Woe to him who builds his house by unrighteousness
> and his upper rooms by injustice,
> who makes his neighbor serve him for nothing,
> and does not give him his wages.

The flip side of the prophetic attacks against economic oppression is the prophecies about the future when forced labor will be abolished. In their visions of a new heaven and a new earth, the prophets portray human beings' special nearness to God and peace in nature, as well as a new quality in human work:

> They shall build houses and inhabit them;
> they shall also plant vineyards and eat their fruit . . . .
> They shall not labor in vain . . ."
>
> (Isa. 65:21,23)

In addition to the enjoyment of the fruit of one's labor, Amos speaks also of the extraordinary productivity in the eschatological new creation (Amos 9:13):

> Behold the days are coming says the LORD,
> when the plowman shall overtake the reaper
> and the treader of grapes him who sows the seed;
> the mountains shall drip sweet wine . . .

The liberation of oppressed people is also a central theme in some New Testament writings. According to Luke's report of Jesus' programmatic sermon in Nazareth, Jesus claimed to be sent "to set at liberty those who are oppressed" (Luke 4:18). The context in Isaiah from which the words were taken (58:6) and Luke's portrayal of Jesus' ministry suggest that Jesus is referring here, at least by implication, to the *economically* downtrodden, proclaiming a visible transformation of economic relations among the people of God.[23] James expresses similar concern for the oppressed. He writes that one essential aspect of "pure and undefiled religion" is to "visit orphans and widows in their affliction" (James 1:28), orphans and widows being the "typical examples . . . for *all* those who suffer distress and oppression."[24]

Alienation and the Experience of Salvation

Humanization of work anticipates the new creation under the conditions of history. This is one good reason why a Christian should strive to humanize work. But a Christian should be concerned about alienating work also because of the intercausal relationship of alienating work with what constitutes the center of the Christian understanding of salvation—the personal relationship between humans and God.

I have argued earlier that alienation from God is the root cause of all other forms of alienation. But this should not lead to the conclusion that these other forms of alienation are inconsequential for the personal relationship between human beings and God. It is too simplistic to postulate a one-way relationship between alienation from God and the forms of alienation caused by this primary alienation. Economic alienation, to stay with our subject, often directly or indirectly causes alienation from God. We read in Exodus, for instance, that the oppressed and exploited Israelite slaves "did not listen to Moses *on account of their* . . . *cruel bondage*" (Exod. 6:9).[25] Economic alienation hindered their believing God and grasping the promises of liberation. (In an important sense, the oppressed *can* be at an "epistemological *dis*advantage!") Similarly, economic alienation of the oppressors can deepen the alienation from God they already experience (because their sin against their fellow human beings is, as such, a sin against God). As the prophetic critique of oppressors indicates, they often sin against God by misusing religious activities to compensate before God for their misdeeds of socioeconomic oppression.[26]

There is thus not only a partial overlapping of alienation from God and alienating work (some forms of alienating work *are* forms of alienation from God) but there is also a causal relationship between them (some forms of alienating work *cause* alienation from God and vice versa), though this causation is asymmetrical, because alienating work is ultimately a consequence of the basic human alienation from God. But the fact that alienating work often causes alienation from God is a good additional reason why a Christian should be concerned about it.

Christian Quietism?

Though Christians generally agree about the responsibility to alleviate suffering when this is possible, they disagree about their responsibility to strive actively for structural and individual change in order to humanize work. Two arguments against this responsibility are most frequently adduced.

The first argument (which has nowadays, fortunately, fallen somewhat into disrepute) rests on the belief in the necessity and virtue of suffering evil. It claims support from the New Testament analogy between the toil of work and the suffering of Christ by construing the injunction to imitate Christ (see 1 Pet. 2:18ff.) to affirm the value of work "*because* of its submissiveness."[27] The cross of Christ is here taken to provide motivation for suffering alienation.

But the cross of Christ does not call Christians to indiscriminate, passive suffering of oppression, but to suffering as one participates in the mission of Christ. A Christian is called patiently to suffer wrong for persistently doing what is right (see 1 Pet. 2:20). There is no virtue in suffering on account of actions that are not right (although even nonvirtuous suffering *may* be beneficial). One cannot therefore start with the necessity of suffering evil and conclude that it is not right to struggle for the humanization of work. One has to start with the question of the rightness of this struggle. If it is not right to struggle for the humanization of work, then Christians are called to suffer alienation; if it is right to struggle for the humanization of work, they are called to suffer in that struggle. So we are back to the original question of whether Christians are required actively to seek to eliminate alienation in work.[28]

Representatives of the second argument against the struggle for humanization of work start with the deap-seated sinfulness of human beings and the consequent permanence of alienation and conclude that all attempts to humanize work are futile. The premise of this argument is correct, but the conclusion is faulty. True, alienation from God, the fundamental form of alienation, inevitably causes alienation in all areas of life—in relations of human beings to themselves, their neighbors, and nature. As I have indicated earlier, since alienation from God will completely disappear only in the new creation, any pre-eschatological attempts to overcome various forms of alienation can succeed only partially.

God's curse after the Fall expresses the fact that alienation is inherent to the human experience of work (Gen. 3:17–19):

> Cursed is the ground because of you;
> in toil you shall eat of it
> all the days of your life;
> thorns and thistles it shall bring forth for you;
> and you shall eat the plants of the field.
> In the sweat of your face
> you shall eat bread
> till you return to the ground. . . .

It would not be correct, however, to hear in this text "a note of pessimism from primordial times."[29] The times have not changed much: modern workers in the high-tech environment of information societies work under the shadow of death and experience struggle and frustration in work no less than their distant forebears (or contemporary neighbors!) did tilling the soil in agrarian societies. The text describes as much some fundamental features of work today as it does those in the primordial time. And its description can be called pessimistic only if we forget that it stands in juxtaposition with the strong affirmation of the nobility of human work (Gen. 2:15). Together, Genesis 2:15 and Genesis 3:17ff. affirm that the inescapable reality of human sin makes work unavoidably an ambiguous reality: it is both a noble expression of human creation in the image of God and a painful testimony to human estrangement from God.

This dual nature of human work not only calls for realism but also prohibits quietism in relation to alienating work. If the doctrine of sin requires a Christian to assert the permanence of alienation in work against all utopian hopes of complete freedom from alienation, belief in God's call of human beings to be co-workers with God requires a Christian to affirm that work *should* be "pleasant, and full of delight, entirely exempt from all trouble and weariness," as Calvin maintained.[30] The presence of the Spirit of the resurrected Christ in the whole of creation, and in particular in those who acknowledge Christ's lordship, gives hope that work also *can* be transformed in ever greater correspondence to this ideal. Christians will thus refuse to accept any given situation as irreformable. They will hope against all hope and strive in the power of the Spirit to make work "full of delight."

FORMS OF ALIENATING WORK

What precisely is the problem with alienating work and what should humanized work look like? As I have already argued, one has to answer this question in terms of the relation of work to human nature. To the extent that work negates human nature, it is alienating; and to the extent that work corresponds to human nature, it is humane. There are different ways work can negate human nature or correspond to it. They represent various dimensions of alienating or humane work. I will try to analyze these in this section.

More than any other part of this study, the discussion of various forms of alienating and humane work needs to be placed in the broader framework of reflection on economic systems. Since theological reflection on economic systems lies beyond the scope of this study (and beyond my competence), I can only state here my assumption that humanized work as I envisage it is best compatible with full-fledged political democracy and fairly consistent (though by no means unbridled) market economy.[31]

Before analyzing individual forms of alienating work, I need to make a few general, explanatory remarks. First, various forms of alienation are not equally problematic. Not allowing the self-directedness of workers, for instance, is obviously a more serious form of alienation than is treating the product of work as a mere means. In concrete situations where various forms of alienation are at play and it is possible to remove one form of alienation only at the expense of the other, one will, therefore, have to live with ambiguities.

Second, not all forms of alienation in work that I will analyze below apply to all types of work. My analysis of the alienation brought about through the contemporary social or technological environment is inapplicable to a single parent doing household work. A single parent manages herself, and the technology at her disposal functions basically as a tool she uses, not as a system into which she has to fit. But some forms of alienation that I will analyze below apply to most types of work. As one does household work, to take the same example, one can treat work and the results of work—in this case, both products (such as a blackberry pie) and states of affairs (such as a clean room)—as mere means, in which case the person doing the household work would be alienated. Though some of what I have to say applies to most of human work, I will not attempt to be comprehensive and reflect on alienation and humanization in all types of work the suggested forms of alienation could apply to, but will concentrate on industrial and information types of work.

My broad definition of work notwithstanding, the analysis of alienation will concentrate on work that is for the most part done as remunerated labor. This is not only because most people still work within employer–employee relationships but also because within such relationships alienation in work is the most acutely felt and the most firmly entrenched. With a little effort, the reader will be able to translate at least part of what I am saying about remunerated labor to other types of work (like household work, voluntary work, work of a self-employed person, etc.).

Third, the forms of alienation in work that I will analyze below do not cover all possible problematic aspects of work. I will not comment, for instance, on the alienation that results from the incomprehensibility and tyranny of the complex social and economic interaction in which the individual's work takes place. The analysis of the causes of this form of alienation and of the ways of overcoming it lies outside the scope of this study. I also leave unexamined many vexatious problems a worker often encounters in work, such as fatigue, unhygienic conditions, dangers, noise, and an ugly environment. Though these problems do fall within the scope of the study, they depend, however, on overcoming the forms of alienation I do discuss and would be easily ameliorated if these forms of alienation were overcome. The same reasoning lies behind my passing over of the problem of exploitation (and frequently resulting destitution) that for so many people in the "Two-Thirds World" is the main form of alienation in work, a form of alienation that often not only violates their human nature but threatens their very lives. Exploitation is certainly an important problem (though one that is not as easily identifiable as it may seem),[32] but it is clearly a derivative one. At its root is the lack of opportunity (both legal and financial) for workers either to take their economic fortunes into their own hands or, if they are employed by others, to participate in significant decision-making.

Autonomy and Development

One of the main forms of alienation in work consists in the lack of self-directedness and opportunity for development in work. Before I investigate this form of alienation as it expresses itself in the relationship of workers to management and technology, I will take a short look through Kant's eyes at Marx's critique of this form of alienation and then indicate briefly how anthropological and pneumatological reflections from previous chapters bear on the discussion.

Legacy of Kant and Marx

One way of analyzing the lack of self-directedness and opportunity for development in work is to think in terms of the inversion of the means and the ends: what should be an end in itself is perverted into a mere means for some other, less noble end. This way of looking at alienation can be expressed also as the reducing of the worker to a mere object.[33] Karl Marx, who built on some aspects of the ethical theory of Kant, is the progenitor of this approach to alienation.[34]

Kant's second formulation of the categorical imperative in *Foundations of the Metaphysics of Morals* reads as follows: "Act so that you treat humanity, whether in your own person or in that of another, always as an end and never as a means only."[35] The idea of humanity as an end in itself functions in Kant's moral philosophy in two related ways: first, humanity as an end in itself sets the *"formal condition of outer freedom"* by imposing a limit on human action; second, it provides "an end which . . . it is a duty to have"[36] by serving as the goal of human action.[37] Respect for humanity requires, negatively, that one not ill-use one's own or another person's humanity by treating it as a mere means; and, positively, that one act in harmony with one's own and another person's humanity by promoting its development.[38]

Marx took over Kant's second formulation of the categorical imperative, modified it to suit his anthropology, and applied it to the industrial world of his time. This is, admittedly, what *I* perceive Marx to have done. He himself, of course, would have resented such an interpretation because, especially after *German Ideology*, he came to consider it useless and ideologically suspect to confront reality with moral imperatives. The whole point of his scientific socialism as opposed to what he called "utopian socialism" was, not to make moral evaluations of capitalist society, but to demonstrate the historical necessity of its transformation into communist society.[39] But what Marx intends to demonstrate to be the result of a historical development can in part be seen as the realization of a modified version of Kant's second formulation of the categorical imperative.

Even if one doubts (as I do) that it is possible to derive "all laws of the will" from the principle of never treating persons as simply means but always as ends,[40] and even if one thinks that Kant gave no good reasons for this principle,[41] one can still follow Marx and look at alienation in work as an inversion of the proper relation between the means and the ends. But a Christian theologian will have to part ways with Marx in two important respects. For one, Marx's analysis was primarily directed against alienating work in industrial production in the capitalist societies. The principle of the inversion of the means and ends can and should be used to analyze alienation also in societies that are not yet and those that are no longer capitalist. Second, and more important, Marx placed his analysis of alienated work in the context of his anthropology and philosophy of history. A theologian will have to place reflection on the inversion of the means and ends in the context of Christian anthropology and pneumatology. This context will provide the legitimacy of the use of the "means–ends model" of ethical discourse in Christian theology.

One of the key words in Marx's understanding of alienating and humanized work is the word "only": the worker (together with her work and its results) should not be treated *only* as means. By implication, so it seems, they can be treated as means *to some extent*. Now, the process and results of work, obviously, can and must be treated as means; they *are* by definition means (though they are by definition not only means). In relation to the process and results of work, the humanization of work will consist in decreasing the worker's experience of work and its results as means.[42] But when one says that *workers* should not be treated as mere means, the quantitative understanding of "only" is inappropriate. Not using other human beings only as means is a question of kind, not degree of abuse.[43] It means one should not reduce them to instruments "through which external forces find enactment"[44] or, in Kant's words, we should relate to them as beings who must "contain in themselves the end of the ... action" we desire them to perform.[45]

Work as Personal Activity

Human beings are created in the image of God in order to have personal fellowship with God.[46] Even though their humanity does not consist in the "power to set an end" (Kant)[47] nor in "free, conscious activity" (Marx),[48] freedom and responsibility are necessary implications of their divinely conferred personhood. A human being should not be compelled to do what he has not freely adopted as his own end. When we exclude the individual's conscious acceptance of the goals we expect that individual to realize, we are treating that person as a mere means. Under the assumption that healthy subsistence is guaranteed, human work should be structured in such a way as not to impede its execution as personal activity.[49]

Since human beings were created to live on earth as God's co-workers in anticipation of the new creation, the Spirit imparted to them various gifts to accomplish that task. These gifts form part of their personality that they are responsible to respect and to develop, both because of the intrinsic value of their personalities as integral parts of the new creation, and because the more they are developed, the better they can anticipate the new creation through their work. Since human beings should strive to image the new creation in the present world as the good that God ultimately desires for them, their personal development is, to use Kantian terminology, a command of morally practical reason, not merely a counsel of technically practical reason.[50] Which of the many aspects of their personalities they should concentrate on developing depends on what gifts they have received from God and what tasks they have been called to accomplish. The duty not to treat other people (and oneself) as a means only and the responsibility to perfect oneself provide important criteria for diagnosing alienation and give direction to efforts to humanize work.

The Old Testament ideal of work as we find it expressed in the legislation on the Year of Jubilee (Lev. 25:8ff.) partly corroborates such an understanding of the humanization of work: in particular, the requirement for work to be a free, personal activity. Every fiftieth year, the Israelites were to set free all the slaves and return alienated land to its original owners. The purpose of this legislation was to "guard against the growth of unhealthy latifundia at the expense of propertyless proletariat."[51] In relation to work, it presupposes as the normative ideal that all individuals (or extended families) should take care of their own needs by working with their own resources, and that they should control production themselves rather than being dominated by others.[52] As one author put it, the point seems to be that every person should "be his own master, freely and joyfully working at his own task."[53]

Worker and Management

At the work place, the most significant relationship workers find themselves in is their relationship to management. To what extent does this relationship foster or hinder workers' self-directedness and personal development? Marx criticized the industrial world of his time by claiming that management makes work "a purely mechanical activity,"[54] stripped of all the characteristics of free art, with devastating results for workers' physical and mental health and development. My concern here is not to discuss whether this is an accurate description of industrial work in the period of early capitalism. Smith, at any rate, agreed with him on this issue.[55] A more significant question is whether or to what extent management reduces workers today to "things."

In the following I will discuss alienation from the perspective of lower-level employees, leaving aside the problem of the self-directedness and development of management. If one were to deal with that problem, one would have to investigate to what extent the behavior of managers is a result of free choices and to what extent they are forced to take certain courses of action by the structures in which they work. To the extent that they are influenced by structures, the analysis of alienation of lower-level employees that follows can be modified and applied to them, too.

Taylorism and Its Problems

The case of persons' being reduced to "things" is the practice of "scientific management," which, in a more or less softened form, still exerts significant influence on management practices in both capitalistic and socialistic economies. The founder of scientific management, F. W. Taylor, formulated one of its basic principles as follows: "All possible brainwork should be removed from the shop and centered in the planning or laying out department."[56] The task of management and the engineering staff is to conceptualize, lay out, and measure in advance the entire work operation, down to its smallest motion. The task of the worker is to renounce all initiative and execute punctiliously the orders given.[57] Taylor's intention was to eliminate the subjective element of the labor process, for he was persuaded that a marked improvement in production would result if each person's work "were as logically ordered as were the actions of the machines."[58] It did occur to Taylor that his management methods turned the worker, as he stated, into "a mere automaton, a wooden man,"[59] but he brushed the disturbing thought aside (with the absurd suggestion that a surgeon is not much better off than

Taylor's worker) because of his concern with maximum efficiency (which he euphemistically called "a fair day's work").

The consistent application of scientific management did result in an increase in productivity, but only to a point. Workers silently rebelled against being turned into machines by decreasing their output.[60] But apart from this economic limitation of scientific management, what exactly is *morally* wrong with it? After all, workers consented (more or less) freely to adhere to the minutely prescribed patterns of action in their work. Were they still used by the management as mere means?

The answer depends on whether the personal nature of human beings requires self-directedness only in accepting or refusing a job or also requires self-directedness in activities *on* the job. It seems to me that the work itself, not only the chance of a job, has to be an *actus personae*. Workers should be able to set goals in their work role and pursue them, or at least be able to identify with the goals that management sets for them.[61] Any action a human being is pressured to do but which she has not made her own goal contradicts her nature as a personal being. Having to do minutely prescribed actions whose relation to the final product a person does not understand hinders him significantly from setting goals for his action and makes it unlikely that he will make the predetermined actions his own goals. Hence when a person works under the conditions prescribed by scientific management he is being reduced to a mere means.[62]

One can imagine a person wanting to execute punctiliously all the minutely prescribed actions and also understanding their relation to the final product. Would the desire to do such work justify doing that work? It would certainly make such work less alienating because it would ensure self-directedness in work. But work under scientific management fails the second test of humanized work because it bars personal development in work. In fact, empirical research seems to confirm Smith's thesis that performing a few simple operations one's whole life has adverse effects on the intellectual life of a person because it indicates that the substantive complexity of work and intellectual flexibility are proportional.[63] So the second problem with Taylorism is that it retrogrades human personality instead of furthering its development.

Taylorism can be effective in forestalling self-directedness and impeding the development of people in a work role because it gives management complete power over the whole structure of work arrangements in the organization. The possession of such power by management does not by itself make the work of lower-level employees alienating. One can imagine that management could use its power to the workers' benefit. But under the pressure of competition, it is highly likely that management will misuse its exclusive power by forcing upon the workers alienating work. Therefore, in order to be able to set goals and prevent the retrogression of their gifts within the work role, workers need to be able to participate in framing the broader conditions of their work. Otherwise they might face an equally unacceptable alternative of being forced either to leave their job or to perform alienating work. If faced with such an alternative, workers would be treated as mere means.[64]

Unfruitful Approaches

Because workers need to participate in decisions about the conditions of their work, it will not suffice to embellish scientific management with programs like job rotation, job enrichment, or job enlargement. These programs are significant insofar as they do increase the substantive complexity of work and broaden the sphere of the worker's responsibility. Empirical studies confirm that these programs do make a difference in job satisfaction of the workers.[65] But they do not go far enough to ensure employees' participation in decisions directly related to their work. For in all these attempts at humanizing work, one thing remains unchanged: the radically asymmetrical relation between the worker and management. One might ask, with a touch of cynicism and exaggeration, whether these programs are not "characterized by a studied pretence of worker 'participation?'"[66]

More radical attempts to ensure self-directedness and a substantive complexity of work propose to pattern all human work according to artistic creation or the professional–client relation. Mechanical mass production is deemed intrinsically alienating. A worker "must considerately hold himself to the solving of each technical problem afresh."[67] Besides unrealistically and radically calling into question the whole modern world of work with its indispensable high productivity, such an approach to overcoming alienation in work puts an unnecessary burden of constant creativity on the worker. In a world populated by people with limited time and expertise at their disposal, there are trivial aspects of any type of work that should be done mechanically so that people's creativity can be directed to other, more significant, aspects of work.

Those with more faith in the inevitability of human progress expect newer developments in the industrial use of microelectronic technology to change the character of human work. The nature of this technology, so they argue, demands restructuring the corporation into a self-managing organization. It is true that information technology is progressively replacing the middle management who "collect, process and pass information up and down the hierarchy" and who supervise the lower-level employees. But does this development warrant the conclusion that "the computer is smashing the pyramid"?[68] Or are computers only destroying the middle link and leaving the vertical hierarchy unaffected?

Information technology is certainly more conducive to the networking style of management than the older industrial technology. Hence it makes self-management more *possible*. But one might argue that it in fact reinforces vertical hierarchies because it facilitates management's control of lower-level employees.[69] Computers can keep track of employees' work much more efficiently than middle management ever could. If used for that purpose, information technology can transmute workers into "glass people" whose every action can be easily monitored. Far from upgrading them into self-managing partners, it can degrade them to closely monitored moilers.[70] Without a conscious effort by management to give up some of their rights and confer them to workers, information technology itself will not eliminate the top-down, authoritarian management style that precludes the workers' participation in decisions about the conditions of their work.

Participation and Development

In order to treat people as ends and not merely as means in accomplishing work, worker participation and development at all levels will need to be encouraged by the creation of work roles that provide enough room for self-directedness and the challenge to use and develop their talents. These two goals of the humanizing of work are closely interrelated but clearly distinct. It is possible to have a bottom-up organization with very low substantive complexity of jobs and a rigid, top-down organization with high substantive complexity of jobs. One must therefore pursue both goals in parallel.

Whether out of respect for people, the desire for increased efficiency in a world where knowledge is becoming the key resource, or both, many corporations today sense a need to restructure work to ensure the participation and development of their people. ServiceMaster—a company that professes to be founded on explicitly Christian principles—is a good example of a company that seeks to promote the personal growth of its employees. In its charter, "People Development Strategy," it states that "people must not be used just as units of production to get the job accomplished" but must be seen "as ends."[71] To treat people as "ends" implies for ServiceMaster an obligation to "help people develop" as persons. Although it views development primarily as the responsibility of each individual, it holds itself "accountable to provide the climate that encourages . . . individual development" by designing positions "to provide stretch experiences that help people grow" and by instituting educational programs designed, not only to improve "work skills," but also teach "conceptual skills and spiritual values."[72] Behind ServiceMaster's commitment to people development lies the theological persuasion that "God created each of us with certain gifts. Therefore, as leaders we should (a) recognize God-given gifts in those we lead and help them develop their gifts to become all that God meant them to be, and (b) recognize that each individual is in the final analysis accountable to be a growing person."[73]

With respect to participation, for many managers today the question is not "*whether* to design organizations for high involvement and self-management, but *how* to do it, and how to do it well."[74] Whether in small steps (increasing autonomy on the job) or in big strides (organizational self-management and self-government[75]) vertical hierarchy seems to be "giving way to the horizontal team where people from different disciplines and perspectives work together on a common goal."[76]

The gradual dismantling of vertical hierarchies does not, however, mean dispensing with leadership. It is a misconception that in self-managing organizations, leadership should help the unit get work underway and then fade into the background. Rather, leadership becomes a "both more important and more demanding undertaking in self-managing units than in traditional organizations."[77] It is more important because of the greater need of integrating a bottom-up organization to keep it operational; it is more demanding because of the intrinsic difficulty of leading by facilitating and because of the constant temptation for the leaders posed by the simplicity and order of an authoritarian organization.

One way to strengthen the tendencies in modern business toward increased participation is to strive to "eliminate the difference between workers, managers, and owners" by compensating workers not only in the form of salary but also in the form of "company stock and profit sharing."[78] Another way would be to encourage the development of the small-scale organization; for, the larger the unit, the more difficult it is for employees to participate in decision-making. The goal, at any rate, should be to make the necessary submission and discipline of work a free act of obedience to their own purposes by all members of the organization.[79]

Worker and Technology

Throughout most of human history, management—to use the modern term—treated workers less as ends in themselves and more as means for their own ends. With the onset of industrialization, a new, and in some respects more enslaving, degradation of workers developed. The masters of work designed the instruments of work that took their place in dominating workers. Admittedly, this is only one side of the story. The same machinery that enslaved laborers to itself contributed over a period of time to their increased material well-being, which was unparalleled in the history of humanity. Nevertheless, the machine could be an efficient helper only by being a harsh master.

The two aspects of alienation in workers' relationship to management that I analyzed in the previous section reappear in the worker's relationship to machinery. Socially and technologically imposed subordination of the worker and stifling of her gifts are two interrelated ways in which she is treated as a means in modern, rationalized economic production. Both stem from the strong tendency of economic organizations striving to maximize efficiency "to reduce the extent to which a profitable operation depends upon discretion of the lower level of employees."[80] Subordination to management is not subordination to a manager as a person (for the authority of a manager is only incidentally located in the person of a particular manager) but to a manager as a representative of a particular *technique* of production. It is this same technique of production that led to the industrial use of technology. Thus one can say that robots represent the result of "industry's logical search for an obedient workforce."[81]

In spite of the close interrelationship of social and technological subordination, however, they clearly represent distinct forms of alienation and require separate treatment. For it is imaginable both that a self-governed organization—say, in striving after profit maximization—would place itself under the tyranny of machinery, and that a tyrannically hierarchical organization would for the same reason encourage workers to play creatively with technology. Another way to make the same point is to draw attention to the fact that "mechanized work" can be done under the pressure of management without the use of machinery, and that "work at the machine" need not be mechanized.[82]

In the following analysis of the relation between workers and technology, my purpose will be to indicate in what ways technology can be alienating and what measures need to be taken in order to humanize work with technology.[83]

Technology, Freedom, and Development

I already analyzed the character of machine production in the industrial era and indicated its effect on human personality: a significant degree of freedom is lost because work is regulated by the action of machines, and there comes about regression in skill level and in the personal development of workers because their activity is reduced to performing a few stupefying operations.[84]

Here I want to note only that the newer developments in information technology do not necessarily improve the relation between worker and technology.[85] True, automation eliminates many repetitive, dangerous, and demeaning jobs by replacing human beings with robots. It also creates an elite of highly skilled specialists. But it would be too hasty to conclude that, on the average, information technology raised the level of skills required. For automation significantly lowers the skill requirements of many traditionally highly skilled jobs. A good example is the effect of CAD (Computer-Aided

Design) systems on the work of an engineer—until recently, among the most creative types of work. They reduce an engineer's work to choosing between different possibilities that a preprogrammed computer suggests. The more pessimistic observers consider the effects of automation on engineers as only one case in point of the general proletarization of white-collar workers that automation induces.[86] Some empirical studies of the effects of technological change suggest that the dominant impact of information technology thus far has been to reduce the substantive complexity of jobs and de-skill workers[87] (though others point out to the Janus-headed nature of the effects of technological change[88]). De-skilling of workers is often accompanied by an increase in workpace and a reduction of work flexibility. When one adds to all these negative effects of information technology the increased capacity for monitoring every aspect of the employees' work that it provides managers,[89] then the fear becomes justified that information technology might be used to pervert human beings into mere means even more ruthlessly and efficiently than mechanical technology ever did.[90]

Legitimacy of Technological Development

Some Christian (and non-Christian) cultural critics tend to consider technology essentially evil, even to the point of accusing anyone fascinated with it of "diabolatry." Such a negative attitude toward technology, which sought biblical support in the story of the Tower of Babel (Gen. 11:1-11), has been so prominent over the past few centuries that it seemed to philosophers like Marx that technological development occurred "in spite of the Bible."[91]

Irrespective of whether they are critics or defenders of technology, contemporary philosophers of science and historians of technology tend to agree that the biblical world-view significantly contributed to modern technological development.[92] It is therefore not surprising that the first pages of Genesis (4:17ff.) give a positive portrayal of incipient technological development at the dawn of history. In the past, this text has been taken to portray the decadence of the Kenites in contrast to the godliness of the Sethites (Gen. 4:25ff.). But if Westermann is right, then it is more correct to view it as making concrete the more general statements about human work found in Genesis 1-3.[93] It describes various forms of human work in their historical development: we read of human beings as farmers (vv. 1f.), architects (v. 17), stock-farmers (v. 20), artists (v. 21), and metalworkers (v. 22). The striking feature of this "history of technological development" is that it does not describe technological advances as ready-made gifts from gods (as does Sumerian mythology) or as jealously guarded possessions of gods (as does Greek mythology), but as results of God's blessing on human ingenuity.[94]

The generally positive attitude toward technological development notwithstanding, the biblical texts show no blind fascination with primal technological achievements. For, like all results of human activity, technology shares the basic human contradiction caused by sin. Technology is not only a noble product of human ingenuity; it is also a dangerous instrument of humans' ignoble ambitions. This is the point the story of the Tower of Babel makes. Without polemicizing against technology as such, it shows the misuse of technology by sinful human beings.

Criteria for Technological Development

Because of its basic ambivalence, technological development needs to be placed under careful theological scrutiny and corresponding practical regulation. In the following brief theological evaluation of technological development, I assume two things about people who live in increasingly technological societies: first, it is possible for them not to be "technicized" to the extent that they became incapacitated to make humane choices in relation to the technology they are creating; and second, within an all-embracing "technological system," changes of human behavior in relation to a particular technology can be significant.[95]

The reflection on human beings and nature in Chapter 5 implies that a particular technology needs to be evaluated in relation to four basic dimensions of human life that correspond with four fundamental needs.[96] First, with respect to the need for God, technological advances should not be used as instruments of rebellion against God or as objects of false worship. As the Tower of Babel (Gen. 11:1-11) suggests, technology can be perverted into a means of human self-aggrandizement. Human beings can use it to assert themselves against the Creator in an attempt to make the dangerous dream of being "like the Most High" come true (see Isa. 14:14). Furthermore, the rapidity of technological development, together with the enticing visions of the future that it promises, can elicit a fascination with technology that borders on idolatry. As J. Ellul states (with his characteristic tendency to generalize the ill effects of technology), the "power of technique, mysterious though scientific . . . is to a technician the abstract idol which gives him a reason for living."[97]

Second, technological development should not be destructive of the nonhuman creation. Industrial production today employs an aggressive technology: its energy sources deplete natural resources and its products and unintended side-effects often cannot be integrated into nature and instead do violence to it. Instead of destroying nature, technology must respect its wholeness.[98] If human work is to be cooperation with nature (which it must be),[99] the instruments of work must cooperate with it, too.

Third, technology should not be developed to foster social disintegration, even less to function as an instrument of social dominion (see Gen. 4:23f.). We should not, however, either think pessimistically that dominion over human beings is an inherent characteristic of technology[100] or expect optimistically that a highly developed technological society would necessarily be a caring society. Rather, we must consciously strive to develop "technology in ways which enhance the quality of human relationships."[101]

Fourth, since the personal character of human beings requires that they act as free agents in relation to technology,[102] technology should not be developed in such way as to hinder self-directedness. After the technological leap from tool to machine, the threat arose that the implacable functioning of machinery would do away with the freedom of the people using it.[103] In the following section I will enquire into how the technologically permeated work process can be made compatible with human self-directedness.

Mastering the Machinery

What can be done, on the one hand, to forestall the use of technology, especially information technology, to enslave human beings, and, on the other hand, to make technology a catalyst of human beings' increased participation and development?

First, workers need to expand their technological know-how. Technological ignorance is alienating for workers who use mechanical technology that is superior to their physical strength and dexterity. It is even more alienating for workers who use information technology that is superior to their mental capacities. Because of the growing rate of knowledge obsolescence,[104] without permanent technological education there is a real danger that the already existing dichotomy between those who understand the functioning of highly complex technology and those who merely push the buttons will evolve into a sharp dualism between a competent minority and an inane majority.[105]

The increase in technological know-how alone, however, will not guarantee free and creative work. The reason for technological enslavement of workers lies not only in their limited knowledge but also in the concrete form of the technology they work with. Competence is of little help if the structure of the work role and the construction of the machinery bars self-directedness and the exercise of one's skills. Therefore, the second condition for overcoming alienation is to design technology that would safeguard freedom and stimulate creativity.

A particular technology might be neutral with respect to the freedom and development of the individual, but it is not neutral as concrete machinery (which is the only way technology can exist, there being no disincarnate technology[106]). "Technologies, all technologies, stem from a social, economic, political, religious and cultural setting. Tools are man's expression of some aspect of his needs, his wants, his hopes and aspirations."[107] If the social and economic goal of work, for instance, is to maximize output, then the technology one uses will have to be designed and constructed to eliminate all factors that could hinder the realization of this goal. In that case, the role of the unpredictable "human factor" will have to be minimized (via fool-proof technology).[108] The result is machinery that objectifies workers by limiting their freedom and stifling their creativity. If, on the other hand, one does not make maximum output the highest priority and considers freedom and creativity in work inherently important goals, it is possible to design machines and to structure work roles to make them compatible with the character of human beings as free and creative agents.[109]

The effects of information technology depend on whether engineers are creative enough and whether owners and managers are willing enough to construct and productively utilize technology in a way that will not be too small for human beings. Through their efforts, information technology can either help overcome the age-old dichotomy between physical and intellectual work or obliterate the distinction between work and idleness; it can either transform work into a creative activity or pervert it into an intensely slavish inactivity.[110]

A third step toward humanization of technology is to trim hypertrophic technology down to a "human scale."[111] Human beings act freely and creatively only if they can understand and control the systems in which they are working. Information technology enables organizations to decentralize into small units and reorganize "production from the huge impersonal production line where man is part of a machine, into a smaller, more independent production unit, where man does what only man can do."[112] Such decentralization, however, will have to be the result of a conscious decision by the management. For it is an illusion to assume "that everything would shrink because certain devices are now so tiny."[113]

Autonomy in Work—To What Extent?

In order to facilitate self-directedness in work and liberate workers from being treated as mere means by management and technology, one will need to strive to facilitate participation, construct human-scale technology that fosters freedom and creativity, and increase the workers' know-how. But even if successfully implemented, these measures will still not make modern workers into *completely* autonomous agents.

Some limitation of individual freedom (above and beyond that which comes from simply being human) is inherent to any work. To work does not mean to be completely autonomous. For work is by definition conditioned by the need it is supposed to supply and by the inherited methods of performing it. It is not completely at our discretion either whether to work or how to work.[114]

In industrialized societies, additional limitations are put on the self-directedness of workers. Their work may be self-directed at the level of individual economic organization, but the organization itself is a part of a larger, highly differentiated national (and even global) economic system consisting of a large number of highly interdependent units. High levels of differentiation and interdependence are consequences of the twin requirements of efficiency and product complexity. Efficiency demands specialization and division of labor. Product complexity, on the other hand, requires integration of specialized skills that are distributed across large segments of society.[115] Whether such integration takes place through the market mechanism or central planning (even of a democratic kind), it is unthinkable without predetermined procedures and rules not subject to change by an individual at his discretion. A certain degree of heteronomy is therefore an integral characteristic of modern economic production. Complete autonomy in work—to the extent that work can be an autonomous activity—is possible only in the enclaves outside industrial production.[116]

The question is whether the necessary lack of autonomy of workers in industrial production makes industrialism inherently alienating. The answer depends on whether the absence of heteronomy at all levels is an indispensable characteristic of humanized work. It seems to me that only freedom in a weaker sense follows from the principle of not treating persons as means—freedom in the sense of being able to set and pursue one's own goals in work (either by creating one's own goals or by identifying with the goals developed by others), not necessarily freedom in the stronger sense of always being able to engage in whatever activities one desires.

Work and the Common Good

Work cannot be humane if it excludes freedom and stifles development of the worker. But free work that fosters one's development is not necessarily humane work. For when one strives for individual freedom and personal development alone, freedom becomes empty— a mere absence of outward regulations for individual behavior— and personal development narcissistic. A person is called to freedom and personal development, but if she is not willing to serve others in love, then, in Pauline terminology, she will be using her freedom "as an opportunity for the flesh" (Gal. 5:13). Without being framed by the concern for the common good, freedom and development, these essential characteristics of humane work, degenerate into forms of alienation: by being free and developing myself I am alienating myself from my true nature as a being-in-communion.

World of Autonomous Individuals

In modern societies it is particularly difficult to think of one's work as a contribution to the common good. It has always, of course, been easier to be served or even to serve oneself than to serve others (see Mark 10:45). The mere fact that the biblical writers needed to exhort people to serve others through work (Eph. 4:28) confirms this truism. Even if one disagrees with Alasdair MacIntyre that "the new dark ages," in which people use moral langauge only to express their emotions and manipulate others, are upon us,[117] one will hardly be able to deny that the structure of economically developed societies and their dominant culture is singularly inimical to the idea of work as a contribution to the common good.

In small and relatively self-contained communities with a low degree of division of labor, the distance between one person's work and the satisfaction of another person's need was small. It was easy to see how one's work contributed to the good of the community. Furthermore, social structure supported one's desire to serve others. In small communities, people experienced "little conflict between their self-interest and the community's public interest . . . because a long term involvement in the community has led them to define their very identity in terms of it."[118] To do good to the community was to do good to oneself, and to harm the community was to harm oneself.

With growing division of labor and complexity of goods and services in modern societies, individuals became increasingly unable to satisfy their increasing needs by themselves and more dependent on the work of others. As Smith correctly perceived over two centuries ago, in order to prosper, individuals in modern societies are forced more than ever to cooperate with one another.[119] At the same time, however, it is equally difficult morally to perceive and perform work as a contribution to a larger common life. For one thing, highly complex economic, technological, and functional interrelations between individuals and social groups are unintelligible to those individuals. A person who often does not even understand how her work relates to the final product is also at a loss to know how her product serves the common good after it enters complex national and international markets.[120]

Moreover, the self-perception of people today does not stimulate them to try to understand the nature of modern economic interdependence and the way their work affects the larger common life. Apart from being involved in small circles of relatives and friends, a person perceives himself as an autonomous individual interacting economically with other autonomous individuals. He owes nothing to the society for his capacities and for what he has acquired through his work; he is an owner of himself, not a part of a larger social whole.[121] Society is only an agglomeration of individuals. What duties can one feel toward congeries of anonymous individuals except to respect each one of them as an autonomous individual? Hardly any. And without a sense of duty toward the larger society, the interest in understanding the communal consequences of one's work dies out.

In modern societies the stress on individual autonomy is accompanied by an equal stress on the pursuit of self-interest. The pursuit of self-interest need not be explicitly egoistic. Smith's famous notion of the "invisible hand" is one way an economic and social theory that stresses pursuit of self-interest pays respect to traditional morality: the "invisible hand" transforms each individual's pursuit of her own interest only into the public good. The best way to work for others is, therefore, to work for oneself. Psychologically, the pursuit of self-interest is justified by the widely accepted notion that "the healthy inner-directed person will really care for others."[122] But whether it is unashamedly egoistic or supposedly indirectly altruistic, economic self-centeredness leaves little place for concern for the common good in one's work. Especially since in market economies the pursuit of self-interest is regularly accompanied by competition for scarce goods, it estranges the worker from the public household and makes it hard for her "to do good work and to be a good citizen at the same time."[123]

Powerful structural and cultural forces in modern societies contribute to the work situation's participation in the general shift from concern for community to preoccupation with self. Whether one works in order to amass goods and enjoy services, to express one's capabilities, or to develop one's potential, the devotion to work is calculated on the benefits it delivers to the individual. A calculating self need not be an isolated self. In modern societies one is forced to place oneself in a large and complex net of cooperative relations. For the majority of lower-level employees, however, cooperation is not an association of people whose wills and energies are directed toward a common purpose, but an interlinking and interworking of people who are indifferent to the goal of the common action.[124] Furthermore, both lower-level employees and managers cooperate, not in order to work in solidarity with one another, guided by a vision of the common good, but exclusively for the benefit cooperation delivers to them as individuals.

Working for One Another

The stress on the pursuit of self-interest in modern societies is at odds with one of the most essential aspects of a Christian theology of work, which insists that one should not leave the well-being of other individuals and the community to the unintended consequences of self-interested activity, but should consciously and directly work for others. Human work has not only personal utility (be it as a means of pecuniary gain or of self-expression), but also moral meaning.

As a rule, the New Testament passages stress that people should work to provide for their own sustenance (2 Thess. 3:12), but also to provide for their needy fellow human beings: they should be doing "honest work" so that they "may be able to give to those in need" (Eph. 4:28).[125] Drawing the implications of Ephesians 4:28, Calvin wrote: "It is not enough when a man can say, 'Oh, I labor, I have my craft,' or 'I have such a trade.' That is not enough. But we must see whether it is good and profitable for the common good, and whether his neighbors may fare the better for it."[126]

One striking New Testament example of conceiving of work as service to others is found in Paul's farewell speech to the Ephesian elders: "In all things I have shown you that by so toiling one must help the weak, remembering the words of the Lord Jesus, how he said, 'It is more blessed to give than to receive'" (Acts 20:35). There is no reason to restrict this admonition only to the church elders who were addressed. The impersonal "one must" may have been used to imply that the injunction applies to *all* Christians.[127] The admonition is radical. For its says more than that every Christian has an obligation to help the poor when she has the means to do so. It charges her not only to help the poor, but explicitly to *labor strenuously (kopiaō) in order to have the means to do so*.

It is a strength of the vocational understanding of work that it conceives of work as service to fellow human beings. Yet since this understanding of work is too static (it sees a person as having only one lifelong calling), it is inapplicable to modern mobile, dynamic societies.[128] A pneumatological understanding of work puts a synchronic and diachronic plurality of jobs or employments in the framework of concern for the common good. The call and empowerment of the Spirit are not the only constitutive features of charisms. So also is their purpose of serving others. The Spirit of God calls and equips people precisely in order to serve their fellow human beings. In the New Testament, charisms are as a rule related to *diakonia* (see 1 Cor. 12:44ff.; 1 Pet. 4:10f.). Their purpose is repeatedly described as the *oikodome* of the community (see 1 Cor. 14:12–26). The reception of charisms obligates one to serve fellow human beings (1 Pet. 4:10), and service constitutes an important criterion of the genuineness of charisms (1 Cor. 14:12ff.). As the Pauline metaphor of the church as the Body of Christ graphically portrays (1 Cor. 12:14–26), charisms link a person to a larger community, a whole in which the charisms of each are a contribution to the good of all.[129]

The anthropological backdrop for the pneumatological understanding of work as a contribution to the common good is the fundamentally *social* nature of human beings. It is a mistake to think of human beings as isolated individuals who, for whatever reason, prefer to live in society but who owe nothing to the society for what they have and what they are. Locke supported his influential notion of an individual as a "proprietor of his own person and capacities owing nothing to society for them"[130] with the theological claim that human beings are "all the Workmanship of one Omnipotent, and infinitely wise Maker; All Servants of one Sovereign Master."[131] Since God made them and since they belong to God, they are both free from each other and equal to each other.[132] This argument is persuasive when used against slavery or serfdom, but it does not establish the anthropological independence of human beings from society. For God does not create each individual human being afresh like Adam out of the dust of the ground; as Jeremy Bentham observed in contradicting Locke, human beings do not come into the world as grown-ups.[133] God creates individuals through other human beings. One becomes a human being and develops as a human being only through social interaction.[134]

Corresponding to the essentially social nature of human beings, the soteriological and eschatological perspectives of the New Testament are communal. Certainly, a person becomes a Christian and lives as a Christian as a free agent, but never as an autonomous being.[135] For the "new man," created "after the likeness of God in true righteousness and holiness" (Eph. 4:24) is a fellowship, not an individual.[136] It is a fellowship of those who were "far off" and those who were "near" (Eph. 2:15ff.), and they form one body, "joined and knit together by every joint with which it is supplied, [which] when each part is working properly, makes bodily growth and upbuilds itself in love" (Eph. 4:13-16). By living as a fellowship of love, Christians image their eschatological destiny and participate in the trinitarian life of the all-loving God.[137] It is, therefore, inadequate to understand the social character of human beings simply as their innate bent to live in a social unit. Animals live in differentiated and interacting units, too. Karl Barth rightly maintained that true humanity is realized only when people live with one another in such a way that they do not live against one another or simply next to one another, but *for* one another.[138]

In order for my work to be humane, I may not work for myself only, forgetting about the good of others or relegating it to the periphery of my consciousness. The good of others must be a goal toward which I am consciously striving. The persuasion that the common good will be realized behind my back, so to speak, as I am busying myself exclusively with myself will not do. It is mistaken because the pursuit of individual self-interest is not sufficient to establish the common good, and because it is also no more capable of making work humane than is unabashed egoism. Because there are no humane people who do not choose to exist for others, there is also no humane work that is not consciously done for others.

I am not, of course, denying that the pursuit of self-interest has, in fact, some beneficial consequences for society as a whole. My point is not to contest the social utility of individual self-interest, but to stress its insufficiency in bringing about the common good. Similarly, I am not proposing self-negating altruism as the only humane alternative to self-seeking egoism. The pursuit of self-interest is inherent to human existence in the world: a being with needs must satisfy these needs, and in order to satisfy them it must be oriented toward their satisfaction. There is an important sense in which a self must be "self-ish" in a world of scarce resources that must be adapted to the needs of the self if the self is to survive. The New Testament recognizes this indirectly by considering provision for one's own sustenance one of the primary purposes of work (2 Thess. 3:10–12).

Individual self-interest can be pursued validly, but it must be accompanied by the pursuit of the good of others. These two pursuits are not in principle mutually exclusive but complementary (though in concrete cases they often conflict). My own good and the good of the whole human family are both included in the *shalom* of the new creation. Therefore, no contradiction is involved when a person "gives himself up" for someone and "loves himself" at the same time (see Eph. 5:25–28). Self-love must be accompanied by self-giving because there is no *shalom* without love for others. If human beings isolate themselves in their work from their close as well as distant neighbors they are denying one of the basic aspects of their humanity.

The horizon of individual concern for the common good is the whole of humanity. Both the universal solidarity of human beings stemming from their common origin and their calling to a common destiny, and the actual universal interdependence of human beings demand such a broad horizon of concern. The individual cannot, of course, be, strictly speaking, held responsible even for the good of her own local community, let alone for the good of the world community, including future generations. But she can keep in mind the good of the world community and attune the pursuit of her own self-interest and the interest of her local community to the good of the world community.[139]

As they work, individuals may not limit their perspective only to local communities, but have to seek the good of all human beings. It is possible, for instance, for work in the armament industry to contribute to the common good on a local level and yet for it to hinder the common good on an international level. Present generations also have to think of the good of the coming generations. Especially in the countries of the Two-Thirds World, younger generations are inheriting the burden of debt, economic failures, and ruined nature from their deceived, unwise, self-seeking, or megalomaniac parents.

Reappropriating the Communal Dimension of Work

To reappropriate the communal dimension of work, it will not be enough to enlighten workers to how their particular work relates to the final product and how the final product contributes to the common good. It will also not suffice to transform the *attitude* of workers toward their work.[140] No doubt, serving others requires the willingness and inner strength that come through the presence of the Spirit of fellowship (see 2 Cor. 13:13). But it is also necessary to create structures that will not foster egoism. On the level of the single corporation, the sharp opposition characteristic of the relationship between management and lower-level employees must be overcome. Individual initiative can be brought to bear on the improvement of the life of all only when structures foster "effective participation in the whole production process, independently of the nature of the services provided."[141]

Since I have already dealt with the issue, I need not here elaborate on the necessity of participation.[142] I will concentrate on the relationship between the market mechanism and concern for the common good. Marx considered the liberal notion that a person serves others by pursuing his own interests unacceptable because, when egoistical individuals "help" each other, "each serves the other in order to serve himself; each makes use of the other, reciprocally, as a means."[143] Moreover, when one elevates private vices to public virtues (as political and economic liberals can be construed to do) one conceals the "mutual plundering"[144] that takes place in market economies. Affirmation of the communal dimension of work requires, in Marx's view, conscious working for others without the mediation of the market. Under the conditions of industrialism, this can occur only through central planning.

But does conscious concern for the common good require a centrally planned economy? Is it incompatible with a market economy? When compared with the state bureaucracy required by planned economies, the market has proven to be a lesser evil. As the bankrupt economies of socialist countries testify, planned economies have economic difficulty satisfying the basic needs of the population. The noble talk about working specifically for others is empty if the economy cannot perform well enough to serve the common good adequately. Worse still, the bureaucracies, with all economic and political power in their hands, have proven more prone to treat people as mere means than small and large capitalists and their managers ever have.

Leaving aside the unsuccessful attempts at establishing an "agreement economy," the only alternative to a planned economy is a market economy directed by a vision of the common good. As I have mentioned earlier, the normative principles implied in the notion of new creation—freedom of individuals, satisfaction of basic needs of all people, and protection of nature from irreparable damage—require the market as an element of a responsible economic system and those principles should, at the same time, be used to establish the parameters for its operation.[145] In the following I will discuss the relationship between individual freedom and the satisfaction of basic needs.[146]

A libertarian concept of a market economy is built around the principle of individual freedom.[147] It defines individual liberty as the basic rule of the economic game and "makes no demands on anyone to accept economic responsibility for others."[148] A specific concern for the satisfaction of the basic needs of all individuals (and also concern for ecological problems) is at odds with its basic tenets. Unlike libertarian philosophy, Christian faith does make demands on people to accept economic responsibility for others. And these demands are not only demands on their generosity. They are demands on them to practice *justice*. Both in the Old and the New Testaments the concept of justice includes concern for the underprivileged (see Matt. 6:1; Ps. 112:9).[149] Paul, for instance, calls the financial help of gentile Christians to the Jerusalem poor "justice" (2 Cor. 9:9). Correspondingly, the mere refusal of the wealthy to aid the poor can be considered a criminal act (Ezek. 16:49).[150]

We should not push aside the biblical language of justice as imprecise because it fails to distinguish what we have learned to keep apart: (procedural) justice and mercy. Instead, we should ask to what extent this terminology requires broadening the concept of human rights and justice. Freedom rights and participation rights are justly held high in all democratic societies. But what about *sustenance* rights? These rights are presently not even generally recognized as rights, let alone consistently respected. The biblical language of justice, together with theological anthropology, imply that we "have a claim on our fellow human beings to social arrangements that ensure that we will be adequately sustained in existence." Therefore, "the deepest answer to the question 'Why care about the poor?' is that if we do not, we are violating the God-given *rights* of other people."[151]

Important as it is, from a Christian perspective, respect for individual liberty will not suffice as a basic rule for the market game. Respect for the right of sustenance of all individuals must be added as a rule that is even more basic than respect for individual liberty. If the market will not behave according to this rule, it is the market that has to go, not the rule. For the basic criterion of the humaneness of an economic system is whether or not it secures lasting justice for the poor.

Incorporating sustenance rights into the market game is only one step in reappropriating the communal dimension of work. That ensures justice. The new creation, however, is a place where love reigns as well. Love is essentially uncoerced. People do not have a legitimate claim on my love that puts me under a moral obligation toward them. There is no general "right" to be loved. One cannot, therefore, implement love by structural change and should not attempt to do it. Love will reign to the extent that each individual opens her heart to others. Since the new creation is a place of love and since love is necessarily a personal activity, there is no virtue in belittling charity. Without it, human relations cannot be humane. As much as we need to recognize that charity is insufficient and structural change is required, we also have to insist that structural change is inadequate and that charity is needed.[152]

If technological developments shorten the work week, people will increasingly need to stand at one another's disposal with their time, energy, and means, even "after hours." In such work for one another they will experience that to give is both more blessed and more humane than to receive (Acts 20:35).

Work as an End in Itself

For the majority of people in modern industrial and information societies (as in all societies that preceded these), work is no end in itself, but a necessary means. They attach no significance to the work of typing, digging, accounting, overseeing, cooking, for example; but see significance in work as *earning* that provides them either with necessities or with luxury, so much so that they tend to define work as a "necessary evil in order to obtain purchasing power over goods and services."[153] If people reflected more on their work experience, they would also recognize the value of significant social contacts the work context provides for them. But in this case, too, work is a means, a means to social contact. For most people, work in itself is not rewarding but is a necessary means to gain access to some desirable goods other than enjoyment of work itself.

One should not slight the monetary and social function of work, especially not in situations where extreme poverty and high unemployment rates deprive many people of these important benefits of work. In economically developed countries, however, people are increasingly dissatisfied with a merely instrumental view of work. Unlike Adam Smith they have come to believe that consumption is not "the sole end and purpose of all production." With Karl Marx they insist that work should be an end in itself.[154]

Work for Work's Sake

Is the notion of "work as an end in itself," however, more than a modern dream, far removed from the real nature of human work? How could work be "an end in itself" when by its very definition it must either be done with the primary goal of meeting the needs of the acting agent or his fellow creatures, or it must be necessary for meeting these needs?[155]

No doubt, an activity cannot lose its instrumental character and still be considered work. But one can choose things for their own sake under the condition of their instrumentality. For instance, there are good instrumental reasons for eating. Much like work, it keeps you alive. But many people could be said to eat not simply to stay alive.[156] As they see it, they eat because they enjoy eating; staying alive is a rather useful byproduct. The same could be true of work. While it cannot objectively be done as an end in itself, subjectively it can be experienced as such.

So far I have stated only that it is possible for work to be an end in itself without contradicting its nature as a primarily instrumental activity. But why should one work for work's sake? Is such work desirable? One could answer the question by saying that work can have its rewards in itself, be "fun," only to the extent that it is perceived as an end in itself. Enjoyment of work is a good reason why one should work for work's sake. But it is not sufficient reason, because then "to be desirable" would be no different than "to be desired." The question is not simply whether "it is fun" when work is fun, but whether work should be an end in itself and, therefore, fun. To find an adequate answer to this question, we need to examine human nature as God desires it to be, not simply human feelings.

If the purpose of human life is either reflection (as in much of philosophical tradition) or worship (as in much of Christian tradition), then work can have only instrumental value. One works in order to keep alive, and one lives in order to think or worship. But if work is a fundamental dimension of human existence,[157] then work cannot have only an instrumental value. If God's purpose for human beings is not only for them to ensure that certain states of affairs come about (the cultivation and preservation of the Garden of Eden) but that *these states of affairs are created through human work* (tilling and keeping), then work cannot be only a means to life whose purpose exists fully in something outside work, but must be considered an aspect of the purpose of life itself. If I am created to work, then I must treat work as something I am created to do and hence (at least partly) treat it as an end in itself.[158]

Therefore, a person cannot live a fully human existence if she refuses to work. This is not the same as saying that she is not fully human if she does not work! For then the aged and ill who can no longer work, and small children who cannot yet work, would not be fully human. Because humanity is exclusively a gift from God, a person can *be* fully human without working, but because God gave him humanity partly in order to work, he cannot *live as* fully human without working. It is, therefore, contrary to the purpose of human life to reduce work to a mere means of subsistence. One should not turn a fundamental aspect of life into a mere means of life. Just as the whole of human life is an end in itself—without, of course, ceasing to be a means to glorify God and benefit the creation—so also must work, as a fundamental dimension of human life, be an end in itself.

The more people experience work as an end in itself, the more humane it will be. If work is to have full human dignity, it must be significant for people *as* work, not merely as a necessary instrument of earning or of socializing; and they must enjoy work. "To work well, one must play at one's work."[159] The Reformers were right in stressing not only that human beings were originally created to work, but also that they were intended to work "without inconvenience" and, "as it were, in play and with the greatest delight."[160]

The stress on "work for work's sake" might seem to play into the hands of the secular fascination with productivity in work and of religious, excessive motivation to work. But taking work, as such, to be pleasurable does not imply being convinced of the all-importance of work. For it is not a contradiction to say that both leisure and work are pleasurable in themselves. In any case, the modern fascination with work does not stem from relating to work as an end in itself. The very opposite is the case. Far from working for work's sake, a modern person striving to produce things most efficiently is not interested in work at all, but in the product of work. The unreachable ideal is to have the product without the work. If, however, work is an end in itself, then the process of working has as much value as the results of work. And when one values work, one will resist pressure to produce frantically and instead take time to delight in work. Human beings are called to achieve something efficiently as well as gifted to enjoy the process of achieving it.

The affirmation of work for work's sake calls into question not only the secular fascination with efficiency but also the traditional religious motivation to work. If one considers work a means of gaining acceptability in God's sight or as proof that one has gained this acceptability (as some Puritans have done, more despite than because of Calvin's teaching), then one will tend not so much to increase productivity (which is what one does when one considers work a means of earning) as to intensify production. For the more one works, the more one will be, or know that one is, pleasing to God.[161] But when one considers work as an end in itself, one will believe that God is pleased not only when human beings work, but also when they delight in their work. One will, therefore, have no reason to shy away from sacrificing the intensity of production for the enjoyment of it.

Work as an End in Itself, and Alienation

It is important to note that experiencing "work as an end in itself" does not, in and of itself, indicate that one is not doing alienating work. Work in which human beings are "at home with themselves" rather than being estranged from themselves is not necessarily humane work, because human beings who are "at home with themselves" need not necessarily be humane. True, most people treat work as an end in itself only when they themselves are treated also as ends in themselves and when they have an opportunity to serve others through their work. Ensuring self-directedness in work, guaranteeing sufficient substantive complexity of work, and making it possible for people to consider their work a contribution to the good of all will foster working for work's sake. But there are people who have learned to find unhealthy pleasure in demeaning kinds of work in which either they themselves or their fellow human beings are being reduced to mere means. In spite of being enjoyable, these kinds of work are objectively alienating.

Just as experiencing work as an end in itself is no guarantee against doing alienating work, so also doing nonalienating work is no guarantee that work will be experienced as an end in itself. The absence of alienation is not a sufficient condition for enjoyment of work. To be enjoyable and humane, work needs to correspond both to human nature (and hence be objectively nonalienating) and to individual gifts and inclinations (and hence be also subjectively nonalienating).

It is a strength of the understanding of work based on the Pauline teaching of charisms that it takes into account the moral character of the context of work and the specific gifts of the individual as well. The vocational understanding of work revolves around God's calling through one's life-setting (*Stand*) and it leaves out of the picture the gifts one has.[162] According to this understanding of work, one can therefore be called to do a particular work irrespective of one's inclinations. The only work one cannot be called to do is immoral work; say, to be a prostitute.

The pneumatological understanding of work, on the other hand, revolves around the individual's gifts (its other foci being God's call and the community's good). One discovers what work God is calling one to do by reflecting on the gifts one has received, not simply by examining one's life-setting. This reflection should, of course, always take place within a given community. A community must recognize my gifts not only as gifts but also as gifts that can be of service to its needs. There is still a need to attune my inclinations to my life-setting. But I cannot be responsible for satisfying any of the community's needs for which I have no gifts. For God does not call a person to do anything for which God does not give her the ability. It is not, therefore, her duty to do whatever morally acceptable work the situation in which she lives might demand of her. It is her privilege to do the kind of work for which God's Spirit has gifted her.

Product as an End in Itself

Because work is a goal-oriented activity, the goal of work is always present in the process of work. Therefore, for work to be an end in itself, the goal must also be partly an end in itself. Marx was at least partly right when he criticized capitalist societies of his day by pointing out that the products are not made because of the immediate relation that the producer has to them but because of the desire for the monetary gain that gives access to other desired products.[163] A worker who is indifferent to what he is making and to how his product is made is estranged from his work. In humane work, a person works in order to make a product, not merely in order to sell it and make a profit.[164] There is, of course, nothing wrong in making products in order to sell them and make a profit. A person who would deny this would either have to accept the absurd idea that each person has to produce everything she needs or the unrealistic idea that only a fully altruistic exchange of services is morally acceptable. My point is that those products a person is making in order to gain access to various goods and services must also be made because they are significant for her as products as well as means. They need not necessarily physically belong to her after they are produced (as Marx thought and thus demanded communal ownership of the means and of the results of production[165]), but she must be able to claim them psychologically. H. de Man rightly spoke of the "instinct" for a finished product and that following this instinct is a presupposition of joy in work.[166] Products have to have an internal significance for producers beyond their purely instrumental utility.

To the extent that modern economies rest on minute division of labor and high complexity of products it is difficult for workers to relate to goods they make as ends in themselves. As Durkheim already observed, if an individual "does not know whither the operations he performs are tending . . . he can only continue to work as a matter of habit."[167] Is Durkheim here describing the inherent qualities of work under the conditions of division of labor? I do not think so. I suggest that it is possible, even when doing such work, to relate to products as ends in themselves if two conditions are met. First, people need to know how their work on the part contributes to the creation of the whole product. In order to gain that knowledge, the technological and operational know-how of workers will need to increase, and the size of production units will need to decrease. Second, people whose individual actions contribute to a product have to experience their work as cooperation. I will be able to claim the joint product as my own and take pride in it only to the extent that I identify with the whole organization that is producing it.

To state that neither work nor the product of work should be a mere means but should also be ends in themselves is to maintain that every good worker goes out of herself and loses herself in her work. Without such "self-forgetfulness," in which the inborn egoism that twists everything into means for our ends loosens its grip on us, there is no true joy in work.[168] The opposition between the self-forgetfulness in work and self-realization through work is only apparent. Just as "everything else" will be added to us when we seek the Kingdom of God (Matt. 6:33), so will self-realization be added to us when we seek good work, when we serve others by self-forgetful, enjoyable work that does not violate our personhood.

Notes

PREFACE

1. See Chapter 4.
2. Because of the fundamental importance of the doctrine of the Trinity of theological reflection, I should here briefly state that I am in sympathy with those theologians who, taking their lead from the Greek church fathers, are attempting to develop a social doctrine of the Trinity. See Moltmann, *Trinität*. For my short reflection on the issue, see Volf, "Kirche," 71ff. For an application of the social doctrine of Trinity to economic issues, see Meeks, *God the Economist*.
3. See "The Oxford Declaration."

INTRODUCTION

1. Pieper, *Leisure*, 21.
2. See Chapter 2.
3. Bellah, *Habits*, 271.
4. Ibid., 288, 289—italics added.
5. See von Nell-Breuning, "Kommentar," 106ff.
6. See *Rerum Novarum*, nos. 1–3; *Quadragesimo Anno*, title page; *Mater et Magistra*, n. 50.
7. *Laborem Exercens*, nos. 1, 3.
8. For a Protestant evaluation of the encyclical *Laborem Exercens*, see Volf, "Work."
9. See van Drimmelen, "Homo Oikumenicus," 66ff.
10. See *The Oxford Conference*, 75–112.

11. See *Labour* and the collection of essays *Will the Future Work? Values for Emerging Patterns of Work and Employment*, ed. H. Davis and D. Gosling (Geneva: WCC, 1985).
12. Stott, *Issues*, xi—italics added.
13. "The Oxford Conference," 23.
14. See "The Oxford Declaration."
15. See, for instance, Johnson, *Grace*; Shelly, *Job*; Raines and Day-Lower, *Modern Work*; Mieth, *Arbeit*; and especially Stott, *Issues*, 154-93.
16. See below, 81-84.
17. Augustine, *Confessions*, XI, 14.
18. *Laborem Exercens*, 3.
19. Ibid., 3f.
20. Simon, *Work*, 1.
21. See Neff, *Work*, 120.
22. For an example of the claim that drudgery is a necessary characteristic of work, see Welty, *Arbeit*, 7.
23. Kluge, *Wörterbuch*, 29; Skok, *Riječnik*, III, 150-51.
24. Simon, *Work*, 20.
25. See Atteslander, "Von Arbeits- zur Tätigkeitsgesellschaft," 125.
26. See below, 196-200. See also Sölle, *Arbeiten*, 83.
27. See *Labour*, 27.
28. Volf, *Zukunft*, 102-4.
29. See below, 134.
30. See Parker, *Work and Leisure*, 25.
31. Heilbroner, *Work*, 12.
32. De Man, *Work*, 67.
33. See below, 177-79.
34. See Marx, *Grundrisse*, 505.
35. Neff, *Work*, 99.
36. See Griffiths, *Wealth*, 110f.
37. See Ricoeur, "Krise," 48; Rinklin, "Bienenfabel," 222.
38. See "Economic Justice," no. 21.
39. See Hartmann, "Ethik," 206ff.
40. So Griffiths, *Wealth*, 66.
41. See Hartmann, "Ethik," 208.
42. Gorbatschow, *Perestroika*, 113.
43. In Marx's view—and in the view of the socialist tradition that is inspired by him—planning was meant to take the place of the market in communism. An even more basic form of alienation in capitalism than exploitation is for him the worker's *lack of control* in economic life as a whole, in spite of the fact that the worker is its primary agent. Detailed planning is necessary, in Marx's view, if workers are to gain full control, not only over the production unit, but over the whole of economic life. Furthermore, whereas in capitalism the economic activity of the individual becomes societal (*gesellschaftlich*) only through the mediation of the impersonal and uncontrollable market (every person works for herself only and the market makes that individualistic work of benefit to the whole society), in communism economic

activity will become consciously societal. Since in an industrial society consciously societal economic activity is possible only through the *organization* of production, planning will be essential in communism. In this respect, Marx's utopia is closer to Comte's (who claimed that the society should be organized "as one single office and one single factory") and, later, Lenin's, than to that of Moses Hess, who thought of economic life in communism in terms of *organism*, not in terms of organization (see Volf, "Das Marxsche Verständnis der Arbeit," 95ff.).

44. So also recently Habermas, "Revolution," 192, 199.
45. See Pomian, "Die Krise," 109.
46. Alperowitz, "Planning," 333.
47. Ibid., 356.
48. For a helpful discussion of market and planned economy from a Christian perspective, see Hay, *Economics*, 144–219.
49. Hay comments: "If the agent is given detailed instructions, and performance is unsatisfactory, it is difficult in an uncertain world to know whether the instructions were at fault, or the conditions were unfavourable, or the agent did not make sufficient effort. The agent has no incentive to exert himself, as he can always shift the blame" (ibid., 204).
50. A dislike for structural change is a consistent feature of all conservative groups—whether they show preference for capitalism or socialism. Conservatives in socialist societies often blame their economic problems on the laziness and irresponsibility of the workers rather than on poor management, mistaken investments, wrong economic policies, or exorbitant arms expenditures. The real reason for such arguments, however, is not a dislike for structural change as such. For they all have no qualms about pleading for structural change in all other camps except their own. The real reason is the conservative desire to preserve intact the social structures in which they themselves function.
51. See Catherwood, "Economics," 4.
52. Quoted by Wogaman, *Economic Debate*, 98.

CHAPTER 1

1. See below, 81–84.
2. See below, 132–33.
3. See Marx, *MEW*, III, 21.
4. See below, 48f.
5. Volf, *I znam*, 99f.
6. See Naisbitt, *Megatrends*, 11ff.
7. See Rasmussen, "Agriculture," 77.
8. See Naisbitt, *Megatrends*, 14.
9. For a description of agricultural work both in the ancient world and in modern mass-production farming, see Kranzberg and Gies, *By the Sweat*, 32–38, 139–48.
10. Rasmussen, "Agriculture," 89.
11. Ibid.
12. Kranzberg and Gies, *By the Sweat*, 9.

13. Xenophon, *Cyropaedia*, VII, 2, 5.
14. Plato, *Republic*, 370 C. Xenophon claims: "he who devotes himself to a very highly specialized line of work is bound to do it in the best possible manner" (Xenophon, *Cyropaedia*, VII, 2, 5).
15. Xenophon, *Cyropaedia*, VII, 2, 5.
16. Kranzberg and Gies, *By the Sweat*, 48.
17. Ure, *Philosophy*, 14.
18. See below, 174–76.
19. See Marx, *Grundrisse*, 374.
20. See Kranzberg and Gies, *By the Sweat*, 92.
21. Smith, *Wealth*, 734.
22. Kranzberg and Gies, *By the Sweat*, 94.
23. Ginzberg, "Work," 73; Pohl, *Divisions*, 44f.
24. See below, 174–76.
25. Quoted by Lyon, *Silicon*, 11–12.
26. Ouellette, *Automation*, 160.
27. See Naisbitt and Aburdene, *Corporation*, 19.
28. Lyon, *Silicon*, 35.
29. Drucker, "Twilight," 100.
30. Kranzberg and Gies, *By the Sweat*, 178.
31. Naisbitt and Aburdene, *Corporation*, 101.
32. Bellah, *Habits*, 127.
33. MacIntyre, *After Virtue*, 122, on "heroic societies."
34. See below, 159.
35. Munroe, Munroe, and Shimmin, "Children's Work," 369.
36. According to some studies, in India, children account for 23 percent of total family income ("All Work," 22).
37. "Children under 15," 116.
38. *Labour*, 8.
39. Studies show that among handicapped, too, there is discrimination along gender lines. The female handicapped seem significantly more "handicapped" than the male (cf. Biondić, *Odgoj*, 58).
40. See Biondić, *Odgoj*, 60ff.
41. See above, 9f.
42. On the problem of reverse discrimination, see Fullinwider, *Discrimination*.
43. See Crosby, "Discrimination," 637ff.
44. Bielby and Baron, "Men and Women," 760.
45. See Naisbitt and Aburdene, *Corporation*, 205.
46. Bielby and Baron, "Men and Women," 761.
47. Quoted by Kranzberg and Gies, *By the Sweat*, 192.
48. Ginzberg, "Work," 73; Erickson, "On Work," 4.
49. Cotton Spinner, "Address to the Public," quoted by Thompson, *Making*, 199.
50. Wogaman, *Economics*, 48.
51. Hay, "Order," 123.
52. See Griffiths, *Morality*, 130ff.

NOTES

53. Hay, "North," 95. See also Hay, "Order," 123.
54. Passmore, *Responsibility*, 43.
55. Ruskin, *Crown*, 49-50.
56. See Walraff, *Ganz Unten*.
57. *The Oxford Conference*, 103.
58. See below 183-85.

CHAPTER 2

1. See Griffiths, *Wealth*, 97-106.
2. By defending Smith against the charge of moral insensitivity and Marx against the charge of totalitarianism I do not want to deny, of course, that each philosopher, in his own way, contributed to such (mis)understanding and (mis)application of his ideas. For a tracing of the connecting lines between Marx's theory and later Marxist versions of totalitarianism, see Kolakowski, *Die Hauptströmungen*.
3. Smith, *Wealth*, 1. See Smith, *Lectures*, 172; Smith, *Draft*, 332.
4. See McNulty, "Smith," 347.
5. See Meek, *Smith*, 15, 18-32.
6. Smith, *Lectures*, 107.
7. Stewart as quoted by Meek, *Smith*, 21—italics added.
8. Reisman went so far as to call Smith "an economic determinist" (*Economics*, 10). But see also his more cautious statements on p. 250. On Marx's view of the relation between economic structure and ideological superstructure, see Volf, "Das Marxsche Verständnis der Arbeit," 80ff.
9. "Whatsoever then he removes out of the State that Nature has provided, and left in it, he hath mixed his *Labour* with it, and joyned to it something that is his own, and thereby makes it his *Property*" (Locke, *Two Treatises*, II, §27).
10. Smith, *Wealth*, 314.
11. Ibid., 315.
12. See Aristotle, *Ethics*, 1140a, 25ff.; 1177b, 15ff.
13. Smith, *Theory*, 153-54.
14. Smith, *Wealth*, 625.
15. Smith, *Lectures*, 338.
16. Smith, *Theory*, 259.
17. Hay, "Order," 88.
18. Smith, *Theory*, 297.
19. Smith, *Wealth*, 33.
20. Ibid., 122.
21. Ibid., 31.
22. See Lindgren, *Smith*, 94, 112.
23. West, *Smith*, 169.
24. See Lyon, *Silicon*, 48, 86.
25. See above, 31.
26. Schumpeter, *History*, 56.
27. Plato, *Republic*, 370B.

28. Smith, *Wealth*, 15.
29. Locke, *Two Treatises*, II, §6.
30. Ibid.
31. MacPherson, *Individualism*, 246.
32. Smith, *Draft*, 328—italics added.
33. Smith, *Wealth*, 11.
34. Ibid., 315.
35. See von Mises, "Liberalismus," 597.
36. Smith, *Wealth*, 13.
37. See Spiegel, "Smith," 103; Arendt, *Vita activa*, 147; Mészáros, *Alienation*, 89f.
38. See Smith, *Wealth*, 13; Smith, *Lectures*, 171.
39. Smith, *Wealth*, 14.
40. Smith, *Lectures*, 253.
41. Smith, *Theory*, as quoted by Novak, *Corporation*, 11.
42. See Bloom, *American Mind*, 165.
43. Smith, *Wealth*, p. 734.
44. Lamb, "Smith," 283. See Smith, *Wealth*, 66, 249.
45. Smith, *Draft*, 327f.
46. Ibid., 328.
47. Ibid., 327.
48. Smith, *Wealth*, 734.
49. Ibid., 735, 740. There is a contradiction in Smith's view on division of labor in *Wealth of Nations*. On the first pages he praises it for increasing skills and stimulating the inventiveness of workers. At the end of the book he blames it for throwing workers into drowsy stupidity. This contradiction defies all attempts at a solution. For an analysis and critique of different attempts to solve the contradiction, see Volf, *Zukunft*, 74f.
50. Smith, *Wealth*, 736f.
51. See ibid., 358, 735; Smith, *Lectures*, 255.
52. Smith, *Wealth*, 734, 740.
53. For a detailed analysis of Marx's understanding of work, see Volf, *Zukunft*, 19–96.
54. See Marx, *Grundrisse*, 7.
55. Marx, *MEW*, XXIII, 193.
56. Ibid.
57. Ibid.
58. Marx, *Grundrisse*, 9.
59. Marx, *MEW*, XXIII, 57.
60. See *MEWEB*, I, 516.
61. See below, 196ff.
62. Marx, *MEW*, XXIII, 192. For arguments against taking this statement as referring to nonhuman nature, see Volf, "Das Marxsche Verständnis der Arbeit," 105, 288, fn. 13.
63. See Marx, *Grundrisse*, 715.
64. Marx, *MEWEB*, I, 517.

NOTES

65. Ibid.
66. See Marx, *MEW*, XXIV/3, 115.
67. Marx, *Grundrisse*, 111.
68. Marx, *MEW*, XXIII, 57.
69. Marx, *Grundrisse*, 505.
70. On relation between work and human nature in Marx, see also below, 200–1.
71. Marx, *MEWEB*, I, 516.
72. Ibid.
73. Marx, *Grundrisse*, 505.
74. Ibid., 156.
75. Ibid., 155.
76. Ibid., 154.
77. In analysis of the division of labor Marx leans on Hegel (see Hegel, *Systeme*, 32, 235).
78. Marx, *Grundrisse*, 204.
79. Marx recognized Smith's concern over the alienating consequences of the division of labor (see Marx, *MEW*, XXIII, 371).
80. Ibid., 371ff.
81. Ibid., 381.
82. Marx, *Grundrisse*, 231.
83. Ibid., 584.
84. Ibid.
85. Ibid.
86. Ibid.
87. Ibid., 599f.
88. Ibid., 313.
89. Ibid.
90. Ibid., 312.
91. Ibid., 313.
92. See Marx, *MEW*, VII, 202.
93. See Marx, *MEWEB*, I, 536.
94. See ibid., 518.
95. Marx, *MEW*, XXV, 784. Marx's concern that we leave nature in better condition for future generations stands in tension with Marx's insistence that the task of human beings is to humanize nature (instead of naturalizing it). See Volf, *Zukunft*, 38.
96. Marx, *Grundrisse*, 387.
97. Ibid., 231.
98. A similar problem can be observed in Smith (see above, 52, 54f) as well as in Fichte's and Hegel's understandings of work (see Volf, *Zukunft*, 75f).
99. Marx, *MEW*, XXIII, 512.
100. See Marx, *Grundrisse*, 593.
101. See ibid.
102. Ibid., 231.
103. See Hegel, *Werke*, VII, §182.
104. Marx, *Grundrisse*, 157.

105. Marx, *Resultate*, 118.
106. Marx, *Grundrisse*, 14.
107. Marx, *MEWEB*, I, 548.
108. Marx, *MEW*, XIX, 21. Cf. Marx, *Grundrisse*, 596.
109. See Marx, *MEW*, XXV, 828.

CHAPTER 3

1. See above, 4–6.
2. The monastic rule *ora et labora* does not imply that worship and work are two alternating aspects of Christian life that cannot be reduced to one another. Work is here a form of ascetic behavior and hence subservient to *vita contemplativa* (see Mieth, *Vita Activa*, 112).
3. Aquinas, *ST*, II–II, Q. 179, A. 2; see Aquinas, *SCG*, III, 135, 13.
4. Aquinas, *ST*, II–II, Q. 182, A. 3, 4.
5. Ibid., A. 3.
6. See ibid., A. 2.
7. See Arendt, *Vita activa*, 20.
8. See ibid., 244ff.
9. See below 138ff. See also Wolterstorff, *Justice*, 151ff.
10. See above Chap. 1.
11. Chenu, *Work*, 4.
12. Clement of Alexandria, *Paedagogus*, 3, 10.
13. Clement of Alexandria, *Salvation*, 16.
14. Ibid.
15. Ibid. The pivotal thesis on the relationship between the human mind and wealth in Clement's treatise represents the exact reversal of Jesus' statement about the relation between the heart and wealth. Jesus claimed: "For where your treasure is, there will your heart be also" (Luke 12:34). Clement, in contrast, says: "For where the mind of man is, there is also his treasure" (*Salvation*, 17). There is, of course, no contradiction between the two statements; both can be true at the same time. But in Clement we encounter a significant shift of emphasis on the inner attitude toward wealth that has had great consequence on the social teaching of the church ever since.
16. Barnabas, xx, 2.
17. Tertullian, "Idolatry," xix. See von Harnack, *Mission*, 300f.
18. Barnabas, xix, 10.
19. Benedict, *Rule*, 48. In non-Christian religions there is a similar stress on the positive effects of work on the souls of the workers. Buddhist monks, for instance, were required to "devote themselves to such tasks in the temple as sweeping and scrubbing, cooking and gathering fuel and so forth. This is based on the idea that even the humblest of activities is a path to enlightenment" (J. Konishi as quoted by Kitagawa, "Work Ethic," 38).
20. Kaiser, "Work," 1016.
21. Ibid.

NOTES

22. See below, 105f.
23. Luther, *WA*, 56, 350 ("... sed ad laborem contra passionem"); see Luther, *WA*, 7, 30. Calvin stressed that by labor the faithful are "stimulated to repentance, and accustom themselves to the mortification of the flesh" (Calvin, *Genesis*, 176).
24. Gillett, *Enterprise*, 58ff.
25. Barth, *CD*, III/4, 531.
26. See *Laborem Exercens*, no. 9.
27. See below, 98f.
28. See Volf, "Doing and Interpreting," 11.
29. For an exception, see Moltmann, "Work."
30. MacIntyre, *After Virtue*, 23.
31. See, for instance, Bienert, *Arbeit*; Richardson, *Work*.
32. See Hengel, "Arbeit," 178.
33. A clear ethical statement in Exod. 22:26–27 provides a good illustration of the difficulty a Christian theologian faces when dealing with the Old Testament ethical instruction. There one reads: "If ever you take your neighbor's garment in pledge, you shall restore it to him before the sun goes down; for that is his only covering, it is his mantle for his body; in what else shall he sleep? And if he cries to me, I will hear, for I am compassionate." On the basis of these verses one can formulate the following socioethical principle (the vagueness of the word "decent" in the last sentence notwithstanding): "The give-and-take of economic life ... may not deprive people of certain fundamental goods such as a cloak to sleep in at night. A person has a right to the material goods she or he needs for a decent existence" (Beversluis, "A Critique," 32). Though the securing of decent existence is an important Christian socioethical principle, a Christian will not accept it simply on the basis of the Old Testament reference in its context cited here. For the same context prescribes the death penalty for sexual perversion (see Nash, "A Reply," 61). Old Testament ethical instructions must be interpreted from within a Christian theological framework.
34. For an analysis of the radical transformation of work in the course of history, see above, 27–35.
35. Moltmann, "Work," 43.
36. See Wolterstorff, "Economics," 15ff.
37. Ibid., 16.
38. See Volf, "On Loving with Hope," 24ff.
39. See Kolakowski, *Gott*, 172; Holmes, *Ethics*, 69ff. For a different view, see Stout, *Ethics*, 13ff, 109ff.
40. Bloom, *American Mind*, 194. On the problematic nature of modern attempts to rationally justify morality, see MacIntyre, *After Virtue*, 36–78.
41. For a pneumatological foundation for respect and tolerance toward non-Christians, see Kasper, "Kirche," 36.
42. Mill, *Utilitarianism*, 234.
43. Contra Arneson, "Meaningful Work," 527.
44. See Rich, *Wirtschaftsethik*, 218.
45. For such a view, see Nash, *Social Justice*, 53ff.
46. Moltmann, *Hope*, 18, speaking of the task of theology as a whole.
47. Bebel, *Die Frau*, I, 410.

48. See below, 94–96.
49. Luther, *WA*, 42, 154.
50. See Langan, "Nature of Work," 121.
51. The need for global ethos and theology was stressed recently by Küng, *Weltethos* and Ambler, *Global Theology*.
52. So Neff, *Work*, 89; see Nielsen, "Needs," 148.
53. Holmes, *Ethics*, 15ff.
54. Marx, *MEW*, III, 52.

CHAPTER 4

1. On that issue, see Berkhof, *Christ*, 189; Dabney, "Die Kenosis"; Gese, "Der Tod," 50; Hoekema, *The Future*, 39–40, 73–75, 274–87.
2. On the relation of *vita activa* and *vita contemplativa*, see above, 70.
3. See above, 71–72.
4. Aquinas, *ST*, II–II, Q. 187, Art. 3. See Luther, *WA*, LVI, 350. For a modern Christian version of this idea, see Schumacher, *Work*, 112ff. In Buddhist tradition, too, work is conceived of as a path to enlightenment (see Kitagawa, "Work Ethic," 38).
5. See Barth, *CD*, III/4, 525.
6. See Williams, "Love and Hope," 26.
7. Under annihilationist presuppositions, Bach's music could not have glorified God directly because it would not have had any intrinsic value. Only the sensations of pleasantness based on a temporary arrangement of matter in the form of human hearing organs would have made Bach's organ music more beautiful and valuable than is the sound of my fingers hitting the computer keyboard.
8. *Evangelism*, 41–42.
9. See Berkhof, *Christ*, 190.
10. *Evangelism*, 42. Cf. Samuel and Sugden, "Evangelism," 208ff.
11. Contra *Laborem Exercens*, no. 26.
12. See Hengel, "Arbeit," 194. Jesus' affirmation of human work is even weaker (though not completely absent) if, as some exegetes suggest, we take "the least of these my brethren" (Matt. 25:40) as referring to Jesus' disciples only (see Gundry, *Matthew*, 514).
13. See Westermann, *Genesis*, 85.
14. On that notion, see Miranda, *Communism*, 12ff.
15. Gundry, "The New Jerusalem," 258, 263.
16. Gundry, *Matthew*, 106, 119.
17. Ibid., 69. In his article on work in early Christianity, Hengel states that a realistic exchatology has its roots "in the realistic preaching of Jesus and is widespread in early Christianity" (Hengel, "Arbeit," 194).
18. See Ratschow, "Eschatologie VIII," 355.
19. Contra Aquinas, *ST*, I–II, Q. 4, Art. 5.
20. Ibid., Art. 7.
21. In his seminal work, *Theology of Hope*, Moltmann has argued persuasively

for a cosmic eschatology as opposed to Bultmann's individualistic and (despite the rhetoric) present-oriented eschatology (Moltmann, *Hope*, 58ff., 133ff.). More recently he developed some of his eschatological ideas in greater detail. In relation to the problem discussed here, see Moltmann, *Der Weg*, 282ff., particularly 286.

22. Cranfield, *Romans*, I, 411–412.
23. Bruce, *Romans*, 170.
24. Calvin, *Institutes*, 989. Some other New Testament passages that affirm continuity between the present and the future orders are: Matt. 19:28; Acts 3:19–21; Rev. 21:24–26.
25. See below, 144f.
26. See below, 143f.
27. See Arendt, *Vita activa*, 26. For a rare theological reflection on the transitoriness and permanence of work see Haughey, *Converting*, 99ff.
28. See below, 128–32.
29. See Rondet, *Arbeit*, 64.
30. The passage refers specifically to the works of steadfastness in faith (see Mounce, *Revelation*, 278), but there is no reason to limit the application of the statement to these works alone.
31. For Protestant examples, see Moltmann, "Work," 38–45, 53–57; Stott, *Issues*, 160f. For Roman Catholic theology, cf. *Laborem Exercens*, nos. 85ff., where John Paul II takes up the theology of work developed in the Pastoral constitution *Gaudium et Spes* (nos. 67ff.) of Vatican II; "Economic Justice," no. 32.
32. See Moltmann, *Creation*, 86; Bienert, *Arbeit*, 45. Differently Preuß, "Arbeit," 614.
33. Luther, *WA*, 15, 373. The last paragraph was taken from Volf, *Zukunft*, 115.
34. See below, 163–68.
35. This is stressed by Honnecker, "Krise," 213.
36. On the "New Jerusalem" as people rather than a place, see Gundry, "The New Jerusalem."
37. See Lochman, *Marx*, 117ff.
38. For the expressions, cf. Kuzmič, "History," 150ff.
39. Moltmann, "Work," 45.
40. See below, 113ff.
41. See above, 79.
42. For examples of the incorporation of creation theology in the present theology of work, see below, Chap. 5 and 6.
43. O'Donovan, *Moral Order*, 55.
44. Heron, *The Holy Spirit*, 154.
45. See below, 143f.
46. See Berkhof, *The Holy Spirit*, 96.
47. On this issue, see Volf, "Materiality."
48. Luther, *WA*, 7, 11, 8–9. Together with *De servo arbitrio* this treatise can most easily be described as a "systematic presentation of his [Luther's] theology" (Ebeling, *Luther*, 212).
49. There is no need to document this statement extensively. I will give only one example. Taking up Luther's distinction between "inward" and "outward man,"

Bultmann writes: when a person becomes a new creation, "outwardly everything remains as before, but inwardly his relation to the world has been radically changed" (Bultmann, "New Testament and Mythology," 20).

50. For a discussion of the differences and similarities between Luther's, Plato's, and Aristotle's talk about inner and outward man, see Jüngel, *Freiheit*, 69ff., 116ff.

51. Luther, *WA*, 7, 21f.

52. Ebeling, *Luther*, 202.

53. See Jüngel, *Freiheit*, 72–73. Calvin seemed to have thought somewhat differently than Luther on the issue: "We should note that the spiritual union which we have with Christ is not a matter of the soul alone, but of the body also, so that we are flesh of his flesh, etc. (Eph. 5:30). The hope of resurrection would be faint, if our union with him were not complete and total like that" (Calvin, *Corinthians*, ad 1 Cor. 6, 15).

54. It should be noted that classical Protestantism did not deny that the full experience of salvation directly affects bodily existence, for it did expect the future resurrection of the body. The point is that the salvation experience does not directly affect human bodily existence *in the present*, i.e., before the consummation.

55. See Schrage, "Heil und Heilung," 200.

56. See Ladd, *A Theology*, 76f.

57. See Moltmann, *Der Weg*, 127. Without knowing the results of modern New Testament studies, Pentecostalists have rightly maintained that by experiencing healing of the body, people became "partakers of the *bodily nature of the kingdom of God*" (Paulk, *Pentecostal Neighbor*, 110—italics mine).

58. Pinnock, "Introduction," 7.

59. Similarly Kasper, "Kirche," 35, with reference to a theology of the world, culture, and politics.

60. Luther, *WA*, 10, I, 311—italics mine. For an early Protestant (and conservative) application of the gifts theme from Romans 12 to the secular and not only the ecclesiastical activities of Christians, see Laurence Chaderton's famous sermon on Romans 12, called "*A fruitful sermon, upon the 3, 4, 5, 6, 7 and 8 verse of the 12 chapter of the epistle of St. Paul to the Romanes*" (Lake, *Puritans*, 28ff.).

61. See, for instance, Mühlen, "Charisma," 168; Lampe, *God*, 202; Taylor, *The Go-Between God*, 26f. For examples from non-Christian tradition, see Plato, who says: "Again, in artificial manufacture, we do not know that a man who has this god for a teacher turns out a brilliant success, whereas he on whom Love has laid no hold is obscure? If Apollo invented archery and medicine and divination, it was under the guidance of Desire and Love; so that he too may be deemed a disciple of Love, as likewise may the Muses in music, Hephæstus in metal-work, Athene in weaving . . ." (*Symposium*, 197Af.). Coomraswamy, following Plato's lead, has suggested a kind of "pneumatological" understanding of work: "So the maker of anything, if he is to be called creator, is at his best the servant of an immanent Genius . . . he is not working of or for himself, but by and for another energy, that of the Immanent Eros, Sanctus Spiritus, the source of all 'gifts'" (Coomaraswamy, "A Figure of Speech," 33).

62. *Documents*, *GS*, n. 38.

63. See, for instance, two contemporary Protestant writers from different seg-

ments of Protestantism, Field and Stephenson, *Just the Job*, 18ff; and Raines and Day-Lower, *Work*, 94ff.
64. See Calvin, *Institutes*, 724f.
65. Luther, *WA*, 34, II, 300.
66. I take it that Luther's use of vocation is not limited to one's standing within the three orders but often equals the person's occupation (contra Bockmühl, "Ethics," 108).
67. Luther, *WA*, 34, II, 306.
68. Wingren, "Beruf," 661.
69. Luther, *WA*, 10, I, 308.
70. Weber, *Ethic*, 80.
71. See above, 70.
72. See Weber, *Ethic*, 282.
73. Calvin, *Institutes*, 725.
74. Luther, *WA*, 10, I, 317.
75. Moltmann, "Work," 47.
76. On Luther's understanding of work as divine service, see Gatzen, "Beruf," 79.
77. See Althaus, *Grundriß*, 80.
78. Luther, *WA*, 42, 640.
79. Calvin claims that God gave human beings vocations because he knew "with what great restlessness human nature flames" (Calvin, *Institutes*, 724). Having a calling from God, a person "of obscure station will lead a private life ungrudgingly so as not to leave the rank in which he has been placed by God" (Calvin, *Institutes*, 725).
80. See Wingren, *Beruf*, 17.
81. Luther, *WA*, 34, II, 307.
82. Wunsch, *Wirtschaftsethik*, 579.
83. See Trilhaas, *Ethik*, 396; Althaus, *Grundriß*, 80.
84. Barrett, *First Corinthians*, 169–70; cf. Brockhaus, *Charisma*, 224; Eckert, "kaleō, ktl.," 599.
85. See Preston, "Vocation," 355: The New Testament term *vocatio* "refer[s] to the call of God in Christ to membership in the community of his people, the 'saints,' and to the qualities of Christian life which this implies."
86. See Wagner, "Berufung," 711.
87. See Käsemann, "Amt," 109–134; Käsemann, "Gottesdienst," 204.
88. On that issue, see Brockhaus, *Charisma*, 220ff.
89. Bruce, *Galatians*, 251.
90. Brockhaus, *Charisma*, 239.
91. For a similar understanding of *charisma*, see Harper, *Let My People*, 100; Mühlen, "Charisma," 161.
92. Küng, *Church*, 246.
93. See Brockhaus, *Charisma*, 170.
94. See Joest, *Dogmatik*, 302.
95. For a similar understanding of charisms in the New Testament, see also Berger, "Charisma," 1105.
96. For an important (but only partial) criticism of Weber's understanding of

charismatic personality and its popular use in Western culture, see Bloom, *American Mind*, 208ff.

97. Schulz, "Charismenlehre," 444.
98. Veenhof, "Charismata," 90.
99. See ibid., 91.
100. The point I am making is not invalidated by the observation that the claim to Spirit's inspiration might have served Israel's kings only as a sacral legitimation of a fundamentally secular power (see von Rad, *Theologie*, 109).
101. See Käsemann, "Amt," 110.
102. Ibid., 118.
103. Moltmann, "Work," 45.
104. For a similar differentiation between calling and mediations within the vocational understanding of work, see Bayer, "Berufung," 142.
105. See below, 163-68.
106. Paul explicates his views on charisms in the context of the understanding of the church as the Body of Christ. He does not derive his views on charisms from this metaphor of the church, but uses the metaphor to illustrate certain aspects of his teaching on charisms.
107. So industrial psychology until recently: see Neff, *Work*, 125.
108. Thomas Aquinas speaks of natural inclinations (caused by divine Providence) to particular employments: "Hæc autem diversificatio hominum in diversis officiis contingit primo ex divina providentia, quæ ita hominum status distribuit . . . secundo etiam ex causis naturalibus, ex quibus contingit, quod in diversis hominibus sund diversæ inclinationes ad diversa officia" (*Quæst. quodliberal*, VII, Art. 17c; cf. Welty, *Arbeit*, 41). As portrayed by Thomas Aquinas, the natural inclinations of different people are as static as Luther's calling and are hence equally ill-suited to modern, dynamic societies.
109. Calvin, *Institutes*, 724.
110. Baxter, as quoted by Weber, *Ethic*, 161.
111. See above, 7-14; Volf, *Zukunft*, 100ff.
112. See Wingren, *Beruf*, 15; Gatzen, *Beruf*, 39ff.
113. Basil, *De Spiritu Sancto*, as quoted by Kern and Congar, "Geist," 87.
114. For the relation between *natura*, *gratia*, and *gloria*, see Moltmann, "Christsein," 626 (though I am not always able to follow Moltmann in the way he determines the relation between *gratia* and *gloria*, and hence also between *natura* and *gratia*).
115. Wolterstorff, "Arts," 467.
116. The claim that "all human activity, including that of work, is captured, permeated and transfigured by the event of salvation" and that "secular reality gains a new—divine—dimension" (Roos, "Work," 103, reporting on French theologies of work) amounts to a dangerous ideology of work if it is understood as an indiscriminate statement about *all* human activity and about the *whole* of secular reality. For some of human activity is beyond salvation and requires abolition (i.e., prostitution), and some of secular reality has demonic dimensions and requires destruction (i.e., chemical weapons).
117. See above, 96-98.

118. See *Documents, GS*, n. 39: "manete caritate eiusque opere."
119. For this interpretation of 1 Cor. 3:15, see Fee, *First Corinthians*, 144.
120. Hebblethwaite, *Hope*, 215.

CHAPTER 5

1. For a discussion of the aspects of work mentioned here as they relate to Marx theory of work, see Volf, *Zukunft*, 119–82. The material presented there and in this chapter complement each other.
2. See below, 127f.
3. Luther, *WA*, 10, I, 310.
4. I am not suggesting, however, that Paul felt merely compelled to preach, for he did his missionary activity in the power of the Spirit.
5. See below, 133–41.
6. Aristotle, *Politics*, 1334a.
7. Arendt, *Vita activa*, 78.
8. Plato, *Republic*, 495E; cf. 611D; cf. Aristotle, *Politics*, 1258b; Xenophon, *Oeconomicus*, IV, 2. 3: "What are called mechanical arts carry a social stigma and are rightly dishonored in our cities. For these arts damage bodies of those who work at them." Aristotle emphasized that a "person living a life of manual toil or as a hired labourer cannot practice the pursuits in which goodness is exercised" (Aristotle, *Politics*, 1278a).
9. Hesiod, *Works*, 175ff.
10. Ibid., 110ff.
11. Carlyle, *Past*, 196.
12. Ibid., 294.
13. See Lochman, "Werk," 110.
14. For a short summary of the debate, see Westermann, *Genesis*, 147–55.
15. On the final syntax, see Schmidt, *Schöpfungsbericht*, 42.
16. Agrell attempted to show that one can interpret the whole of Genesis 2–3 partly as a response to the problem of human work (cf. Agrell, *Work*, 8ff.).
17. Hengel, "Arbeit," 179.
18. See above 70, 105f. Calvin maintained that human beings "were created for action," "to employ themselves in some work, and not to lie down in inactivity and idleness" (Calvin, *Genesis*, 175, 125). According to Luther, too, human beings are "born to work as birds are born to fly" (Luther, *WA*, 17, I, 23). On Luther's understanding of the relationship between human nature and work, see Bayer, "Tu dich auf!" 70.
19. Habermas, *Erkenntnis*, 80. See also Scheler, "Arbeit," 174.
20. Von Rad, *Genesis*, 95.
21. The two perspectives are, of course, interdependent. The way others perceive me flows into the way I see myself, and the way I see myself influences the way I am perceived by others. On the need for attunement of internal and external perspectives in the self-definition of individuals and social units, see Welker, "Towards a World Theology," 441ff.

22. See Bellah, *Habits*, 56.

23. Hunger for self-realization explains, not only continued stress on work, but also the tendency to transpose the work mentality into leisure experience (see Johnston, *Play*, 20f.).

24. Naisbitt and Aburdene, *Corporation*, 47.

25. See above, 51f.

26. See Volf, *Zukunft*, 31ff, 73ff.

27. Neff, *Work*, 159. On the basis of investigations that show that there is a significant correlation between low substantive complexity of work and low ideational flexibility—a very stable personality trait—Kohn and Schooler have concluded that job conditions are molding personality ("Substantive Complexity of Work," 122).

28. See Branda, *Work*, 141.

29. See Krusche, *Das Wirken*, 95-125.

30. See above, 112.

31. As it is normally understood in the psychological literature, "self-actualization as an ethical goal is individualism writ large, with a pseudo-biological sanction" (Smith, "Needs," 138).

32. Fitzgerald, "Needs," 49.

33. Hegel, *Werke*, VII, 124.

34. Nietzsche, *Moral, KGW*, VI, 2, 293; cf. Nehamas, "How One Becomes What One Is."

35. On this issue, see above, 57f. For a detailed argument in favor of the interpretation of the relationship between work and humanness in Marx given in the following paragraph, see Volf, "Das Marxsche Verständnis der Arbeit," 8-169.

36. See Marx, *MEW* III, 21, 85. For a similar view by Hegel, see *Werke*, XXIII, 336.

37. Marx, *MEW* III, 21—italics added.

38. Ibid., XXVI/3, 482. One wonders whether Marx's view of the self-production of the human race is not merely a socialistic reflection of the hard-core capitalist ideal expressed in Bounderby's boast about his economic success: "Here I am . . . and nobody to thank for my being here, but myself" (Dickens, *Hard Times*, 12).

39. Marx, *MEWEB* I, 516.

40. Differently, Sölle, who writes that work is one way to *become* (not merely to live as!) the image of God (Sölle, *Arbeiten*, 110).

41. For an explication and justification of this anthropological thesis, see Dalferth and Jüngel, "Person," 70; Moltmann, *Creation*, 226ff.

42. Neulinger, *Leisure*, 31.

43. See on that Parker, *Work and Leisure*, and Neulinger, *Leisure*, 3ff.

44. On definition of work, see above, 7-14.

45. For further elaboration of my definition of leisurely activity, see Volf, *Zukunft*, 159-61.

46. Contra de Man, *Joy in Work*, 19.

47. See below, 195-200.

48. "America Runs Out of Time," 53.

49. Von Nell-Breuning, *Arbeitet der Mensch zu viel?* 98.

NOTES

50. Neulinger, *Leisure*, 155ff.
51. "America Runs Out of Time," 52. Analysis shows that "employed American adults had, as a total population, *no* net gain in their leisure over the 30-year period following World War II" (Kraus, "New Leisure," 12).
52. See Ellul, *System*, 314.
53. So also Hawtrey, "Work and Leisure," 16.
54. Aristotle, *Politics*, 1337b 30.
55. See below, 152–54.
56. See Volf, *Zukunft*, 161ff.
57. See Käsemann, "Gottesdienst," 200f.
58. Though I acknowledge that the whole life of a Christian should be worship, for lack of better term, I will follow the traditional Christian terminology and use "worship" to describe a person's private or liturgical communion with God.
59. On this eschatological hope, see Gundry, "New Jerusalem," 262. On indwelling of God in the new creation, see Moltmann, *Der Weg*, 353.
60. The phrase comes from Plato (*Laws*, 653D) and has its original home in a quite different theological milieu than I am placing it in.
61. See above, 70, 90. See also Zimmerli, "Arbeit," 50; Bienert, *Arbeit*, 23.
62. Aristotle, *Ethics*, 1177b, 5.
63. See above, 70.
64. See Volf, "Doing and Interpreting," 12–14.
65. M. Mead in Johnston, *Play*, 85.
66. Pieper, *Leisure*, 38.
67. See above, 27f.
68. See Schweizer, *Geist*, 128. To make my point, I do not need to claim that the New Testament knows of no hierarchical order whatsoever between charisms; say between the apostolate and glossolalia (as Schweizer seems to imply). It suffices to establish that the New Testament knows of no hierarchical ordering between charisms that relate to "secular" realities (such as that of a deacon) and to "spiritual realities" (such as that of a prophet).
69. Wolterstorff, *Justice*, 147.
70. Augustine, *De Civitate*, xix, 19.
71. Already Fichte has demanded that leisure be recognized as a basic human right (Fichte, *Nachgelassene Werke*, II, 535f).
72. Parker, *Work and Leisure*, 114.
73. These studies, claims Hawtrey, seem to indicate that "the less religious a person is, the more he or she is inclined to disregard the intrinsic value of work and be more leisure-oriented . . ." It was also "found that those who professed to be atheists or to have no religion at all tend to be less concerned about being useful, took less pride in work, wanted more holidays and felt exploited more often" (Hawtrey, "Work and Leisure," 17). My point here is not to defend a particular influence either of religion or of atheism on work, but to indicate that such influence does exist.
74. Heschel, *Sabbath*, 29.
75. Underhill, *Worship*, 18.
76. Marx, *MEWEB* I, 512.

77. Birch as quoted by Naisbitt, *Megatrends*, 15.
78. See above, 41–42.
79. To give an influential contemporary example, John Paul II writes that Genesis 1:28 expresses the "very deepest essence" of work (*Laborem Exercens*, no. 4). The first chapters of the encyclical can be seen as an extended explication of this verse (see Volf, "Work," 67f.).
80. For Bacon's views on subduing nature, see Leiss, *The Domination of Nature*, 48ff.
81. *Laborem Exercens*, no. 4—italics added. Cf. Jacob, *Genesis*, 61.
82. Descartes, *Discourse*, 49.
83. Descartes, *Meditations*, 132–33.
84. See Zimmerli, "Mensch," 149; Steck, *Welt*, 68.
85. See Moltmann, *Creation*, 244ff.
86. On the presence of the Spirit in creation, see Moltmann, *Creation*; Schweizer, *Geist*, 25ff.; Congar, *Der Heilige Geist*, 311ff.
87. Calvin, *Institutes*, 138.
88. See Koch, "Gestaltet die Erde," 36. On God's rejoicing in creation for its own sake, see McPherson, "Ecological Theology," 237ff.
89. See above, 58, 62f.
90. See Sölle, *Arbeiten*, 12.
91. See Feuerbach, *Werke*, VI, 134.
92. See Westermann, *Genesis*, 221.
93. Luther, *WA*, 42, 77. Cf. Calvin's statement: "Let him who possesses a field, so partake of its yearly fruits, that he may not suffer the ground to be injured by his negligence; but let him endeavor to hand it down to posterity as he received it, or even better cultivated. . . . [L]et every one regard himself as the steward of God in all things which he possesses. Then he will neither conduct himself dissolutely, nor corrupt by abuse those things which God requires to be preserved" (Calvin, *Genesis*, 125).
94. Wolterstorff, "Arts," 466, of the work of an artist. Cf. Jensen, "Materialismus," 253; Tödt, "Die Ambivalenz," 28; Altner, "Technisch-wissenschaftliche Welt," 96.
95. See Caterwood, "Technology," 128.
96. Wilckens, *Römer*, II, 158; Balz, *Heilsvertrauen*, 51.
97. Liedke, *Im Bauch*, 137.
98. See Calvin, *Genesis*, 125.
99. Zimmerli, "Arbeit," 43.
100. Ibid.; see also Westermann, *Genesis*, 159.
101. Koch, "Gestaltet die Erde," 28.
102. See Westermann, *Schöpung*, 76ff.
103. See *The Oxford Conference*, 100; *Rights*, 12f. See also Locke, *Two Treatises*, I, §24ff.
104. Bacon, *New Organon*, 247f., as quoted by Leiss, *Dominion*, 49.
105. Bacon, *New Organon*, 115, as quoted by Leiss, *Dominion*, 50.
106. Luther, *WA*, 49f.

107. See *Altra-Hasis*, 43ff.
108. Barth, *CD*, III/4, 525.
109. See Preuß, "Arbeit," 616; von Harnack, *Mission*, 198.
110. With the term "product-needs," I refer to human needs that must be satisfied with material products.
111. Smith, *Lectures*, 160.
112. See Kant, *Urteilskraft*, §83; Hegel, *Werke*, VII, §185 Zusatz.
113. For some contrary examples, see Sahlins, *Stone-Age*, 1-39.
114. See Ignatieff, *Needs*, 141.
115. Already Hegel reflected on the so-called artificial needs created by the producers in order to keep the production going (Hegel, *Werke*, VII, §190, Zusatz; see also Marx, *Grundrisse*, 14). In our time the "new left" (Habermas, Marcuse, Fromm) have elaborated on these themes in their criticism of capitalism. For a theological reflection on this problem, see Meeks, *God the Economist*, 157ff.
116. Plato, *Republic*, 561.
117. See Nielsen, "Needs," 142f.
118. Ignatieff, *Needs*, 11.
119. For further development of this theme, see Volf, *Zukunft*, 152ff.
120. Augustine, *Confessions*, III, 1.
121. Pieper, *Leisure*, 59.
122. See above, 143f.
123. Maslow, *Personality*, 99.
124. For further reflection on working for the good of all humanity, see below, 186-95.
125. See above, 130.
126. Jüngel, "Eschatologie," §10.1. The last paragraph was taken from Volf, *Zukunft*, 157f.
127. For a different opinion, cf. Langan, "Nature of Work," 130.
128. See above, 37f.
129. See Sölle, *Arbeiten*, 135.
130. W. Temple, as quoted by Stott, *Issues*, 163.
131. *The Oxford Conference*, 91f.
132. See above, 108f.

CHAPTER 6

1. I elaborate here on some basic ideas about alienation and humanization of work developed in my critical dialogue with Marx's concept of alienation (Volf, *Zukunft*, 126-40).
2. See Hall, *Work*, 92ff.
3. See, for instance, Jahoda, *Arbeit*, 106.
4. Kohn, "Occupational Structure," 86f.—italics added.
5. De Man, *Work*, 11.
6. Calvin of Adam's work in the Garden of Eden (Calvin, *Genesis*, 125).

7. This should not be taken to imply that workers' satisfaction does not sometimes increase with the increase in substantive complexity of work (see Friedmann, *Work*, 14).
8. Kono, *Japanese Enterprise*, 332.
9. Hall, *Work*, 92.
10. Marx, *MEW* III, 52.
11. Anthony, *Ideology*, 227.
12. On joy in spite of alienation, see Volf, *Zukunft*, 141.
13. See Seeman, "Alienation Motif," 172f.
14. See above, 128–33.
15. See below, 195ff.
16. Se below, 196f.
17. The "fit" hypothesis stresses the need for fitting worker characteristics with job characteristics (see Hall, *Work*, 96ff.).
18. See above, 53–55.
19. See above, 152–54.
20. Marx, *MEWEB* 1, 521. On the problematic nature of Marx's tracing all alienations back to economic alienation, see Volf, *Zukunft*, 46f., 53f.
21. Marx, *MEWEB* 1, 521.
22. See above, 99–102.
23. For a somewhat stronger formulation of the same thesis, see Yoder, *Politics*, 39.
24. Davids, *James*, 103—italics added.
25. It is interesting to note that the Egyptians' strategy to prevent the Israelites from listening to Moses was to increase their labor: "Let heavier work be laid upon the men that they may labor at it and pay no regard to lying words" (Exod. 5:9).
26. See Kraus, *Theologische Religionskritik*, 251.
27. Heilbroner, *Work*, 15. In recent theological literature this analogy was given a positive, liberating meaning. See Moltmann, "Work," 42ff.; *Laborem Exercens*, no. 27; Mieth, *Arbeit*, 36ff.
28. See in relation to this also above, 81–84.
29. Günkel, *Genesis*, 23.
30. Calvin, *Genesis*, 125.
31. See above, 14–21 for short reflections on the issue. The notion of humanized work I am suggesting is most compatible with a kind of economic system implicit in Hay's critique of market and planned economies, with which I, on the whole, agree (see *Economics*). After having lived through it, I am less persuaded of the virtues of Yugoslav experiment than Hay is (*Economics*, 212, 219).
32. On exploitation, see above, 40f.
33. For a person not to be treated "only as means" is for Kant synonymous with a person's not being treated as a thing (Kant, *Grundlegung*, A428f.).
34. See above, 58–61.
35. Kant, *Grundlegung*, A429.
36. Kant, *Metaphysic*, A380.
37. Potter, "Kant on Ends," 78; see Korsgaard, "Humanity," 185.
38. See Kant, *Grundlegung*, A429f.

NOTES

39. See on this issue Volf, "Das Mauxsche Verständnis der Arbeit," 78ff.
40. Kant, *Grundlegung*, A429.
41. See MacIntyre, *After Virtue*, 44f.
42. See below, 195–201.
43. Contra Acton, *Kant's Moral Philosophy*, 37.
44. Scruton, *Kant*, 71.
45. Kant, *Grundlegung*, A430.
46. See above, 132f.
47. Kant, *Metaphysic*, A392; cf. Kant, *Grundlegung*, A437. For problems with identifying rational nature with humanity, see O'Donovan, *Moral Order*, 47f.
48. Marx, *MEWEB* I, 516.
49. On work as an *actus personae*, see *Laborem Exercens*, no. 6. According to my analysis, the statement that "no one may 'own' the labor of another person" (Meeks, *God the Economist*, 147) is correct, only if ownership implies *ius abutendi*.
50. See Kant, *Metaphysic*, A387.
51. North, *Sociology*, 163.
52. See Hay, "Order," 90.
53. Solomon, *Lösung*, 45.
54. Marx, *Grundrisse*, 204.
55. See above, 53–55.
56. Taylor, *Management*, 98.
57. Taylor relates the frequent reaction of workers "when they first come under this system": "Why, I am not allowed to think or move without someone interfering or doing it for me!" (Ibid., 125).
58. Kranzberg and Gies, *By the Sweat*, 153.
59. Taylor, *Principles*, 125.
60. On workers' resistance to being treated like machines, see Ginzberg, "Work," 73.
61. As I see it, the problem of alienation is not so much the problem of powerlessness understood as a (perceived or actual) inability of a person to "determine the occurrences of the outcomes . . . he seeks" (Seeman, as quoted by Kohn, "Occupational Structure," 86) as it is the problem of the lack of self-directedness in setting the goals of one's actions and pursuing them.
62. This does not make the automatic actions of human beings subhuman, provided that they will to do these actions, that these actions do not hinder their personal development, and they are necessitated by the pursuit of some other morally noble end.
63. See Kohn and Schooler, "Substantive Complexity of Work," 103ff.; Spenner, "Deciphering Prometheus," 827.
64. Given my way of arguing in favor of self-directedness in work, this argument for workers' participation in management is valid only if it can be expected that managers will not respect workers' self-directedness. And this is indeed what is to be expected in the economic framework characterized by enforced profit maximization. To the extent that managers do respect workers' self-directedness—which happens sometimes, too—my argument is not only invalid but proves also to be unnecessary.
65. Jahoda, *Arbeit*, 129f.

66. Braverman, *Labor*, 39.
67. Ryan, "Humanistic Work," 15.
68. Naisbitt and Aburdene, *Corporation*, 12, 5.
69. See Lyon, *Silicon*, 41, 107; Collste, "Work Ethic," 95.
70. See Evans, "Arbeitnehmer und Arbeitsplatz," 189f. There are, however, important ways information technology can contribute to the increase of the individual employees' freedom, even without calling into question the vertical hierarchy of the organization. Because work can be effectively controlled without middle management, employees can, for instance, have flexible work schedules and work at home. Moreover, if by increasing productivity, information technology contributes to a general reduction of work time, then it can create new possibilities for *autonomous work*—work outside of one's formal employment done to supply one's own needs or the needs of fellow creatures.
71. "People Development Strategy," 1.
72. Ibid., 3, 2, Attachment A, 1.
73. Ibid., Attachment A, 1.
74. Hackman, "The Psychology of Self-Management," 89.
75. In a self-*managing* unit, "members have responsibility not only for executing the task but also for monitoring and managing their own performance," and in a self-*governing* unit, they "decide what is to be done, structure the unit and its context, manage their own performance, and actually carry out the work" (Ibid., 92f.).
76. Naisbitt and Aburdene, *Corporation*, 39.
77. Hackman, "The Psychology of Self-Management," 119.
78. Naisbitt and Aburdene, *Corporation*, 54; cf. Hay, *Economics*, 173.
79. See Heilbroner, *Work*, 24.
80. Hill, *Competition*, 122.
81. Cooley, *Architect*, 68.
82. See De Man, *Work*, 110ff.
83. Earlier I have given reasons why the lack of self-directedness and substantive complexity in work are alienating, so I need not reiterate them here (see above, 170–73).
84. See above, 39.
85. Friedmann rightly considered it utopian to think that automation automatically brings about the best conditions in the working place (Friedmann, *Work*, 120).
86. See Cooley, *Architect*, 38.
87. See Lyon, *Silicon*, 37; Janzen, "Auswirkungen," 29. Sociologists often bemoan the Janus-headed nature of the effects of technological change.
88. Spenner concludes that "there is as yet no verdict on the upgrading, downgrading, and little-net-effect hypotheses" (Spenner, "Deciphering Prometheus," 825).
89. See above, 177.
90. See Evans, "Arbeitnehmer und Arbeitsplatz," 187. On the negative effect of automation, see also Erickson, "On Work," 4ff.
91. Marx, *MEW*, XXIII, 194.

92. This does not imply that the biblical world-view should receive all the blame for the destructive use of modern technology (see above, 146–48).

93. See Westermann, *Genesis*, 56ff., 321ff.; Wolff, *Anthropologie*, 122; Preuß, "Arbeit," 615.

94. See Westermann, *Genesis*, 62. One of the early Christian defenders of technological development was Origen. Answering a charge of Celsus that it does not seem plausible that God would provide for savage animals but have human beings labor and suffer to earn a "scantly and toilsome existence"—a charge that presupposes a typical Greek negative attitude toward work—Origen says that God, "wishing to exercise the human understanding in all countries . . . created man a being full of wants, in order that by virtue of his very needy condition he might be compelled to be the inventor of arts, some of which minister to his subsistence, and others to his protection" (Origen, *Celsus*, IV, 76).

95. For different assumptions, see Ellul, *System*, 310ff.

96. See above, 152–54.

97. Ellul, *Society*, 144f.

98. See Chaw, "Technology," 18f.; McPherson, "Ecological Theology," 239.

99. See above, 144–46.

100. See Ellul, *System*, 5.

101. Lyon, *Society*, 27.

102. See *Laborem Exercens*, no. 5.

103. See Ellul, *System*, 17.

104. See Cooley, *Architect*, 26.

105. See King, "Einleitung," 38f.

106. See McPherson, "Ecological Theology," 239.

107. Shallis, *The Silicon Idol*, 94. "The entire innovation process, from basic research to the marketing and use of a new technology is conditioned by such factors as profit motive, prestige, national military goals, and social and economic policies" (Norman, *God That Limps*, 185).

108. Hill, *Competition*, 103.

109. See Brakelmann, "Arbeit," 115; Lyon, *Silicon*, 21.

110. See Jonas, *Verantwortung*, 351f. It depends on the construction of machinery whether it is true that "the best way of demechanizing work is to hand it over to a machine" (De Man, *Work*, 112).

111. Schumacher, *Work*, 23; cf. Schumacher, *Small*, 52ff.; Illich, *Selbstbegrenzung*, 30ff.

112. Catherwood, "Technology," 137.

113. Ellul, *System*, 2.

114. See above, 11f.

115. Gorz, *Wege ins Paradies*, 81.

116. See Auer, *Umweltethik*, 170.

117. MacIntyre, *After Virtue*, 263. For a critique, see Stout, *Ethics*, 191ff.

118. Bellah, *Habits*, 175.

119. See above, 52.

120. See De Man, *Work*, 70.

121. See MacPherson, *Individualism*, 3. Economic development and individualistic social philosophy and practice are separable. The countries of Pacific Rim especially have managed to reach high levels of economic development without succumbing to the kind of individualism that has accompanied and fostered economic development in the West.

122. Bloom, *American Mind*, 178. Bloom continues: "To which I can only respond: If you can believe that, you can believe anything."

123. Bellah, *Habits*, 197.

124. Hannah Arendt (*Viva Activa*, 112) denies that cooperation can exist under a condition of advanced division of labor because cooperation presupposes a difference between cooperating individuals, whereas in industrial work individuals are interchangeable. But cooperation requires only that *tasks* of individuals differ, irrespective of whether each individual can in fact do all the tasks. Eighteen workers each of whom produces one-eighteenth of a pin—to take the famous example Adam Smith uses—are cooperating. In fact, the level of cooperation is proportionate to the level of division of labor. In industrial and information societies one can call in question only the nature of cooperation, not the fact of it (see Friedmann, *Work*, 69ff.).

125. See Volf, "Work," 77.

126. Calvin, *Ephesians*, 457.

127. See Agrell, *Work*, 137f.

128. See above, 108f.

129. Thus, Bellah of calling (*Habits*, 66). Bellah has suggested that the reconstitution of the social world he is pleading for demands a reappropriation of the idea of "calling" (66ff., 287ff.). If he is not suggesting that one should have only one calling that defines one, as he sometimes seems to (see 69, 118, and especially his reference to MacIntyre, *After Virtue*, Chap. 10), then his valid concern would be met equally well by my pneumatological understanding of work as by his vocational understanding of work.

130. MacPherson, *Individualism*, 255.

131. Locke, *Two Treatises*, II, §6; cf. I, §53.

132. See Lasalett, "Introduction," 92.

133. In fairness to Locke, he considered that children are born "to equality" but not "in it" (Locke, *Two Treatises*, II, §55). This is one of very few instances in which age and development are, for Locke, relevant to human relations (see Lasalett, "Introduction," 94), but it has no systematic consequences for his anthropology and social theory.

134. On the relation between individual and community, see Niebuhr, *The Children of Light*, Chap. 2.

135. See Volf, "Kirche," 67, fn. 62.

136. See Barth, *Ephesians*, 537.

137. See Volf, "Kirche," 70ff.

138. See Barth, *CD*, III/2, 223-85; Pannenberg, *Anthropologie*, 516. Gandhi has emphasized that "the duty of renunciation differentiates mankind from the beast" (Gandhi, *Labour*, 14).

139. See Hartmann, "Ethik," 201.

NOTES

140. See Rich, "Arbeit," 16f.
141. *Laborem Exercens*, no. 13.
142. See above, 172ff.
143. Marx, *Grundrisse*, 155.
144. Marx, *MEWEB*, I, 460.
145. See above, 14–21.
146. On work and nature, see above, 141–48.
147. See Friedman, *Capitalism*, 12.
148. Griffiths, *Wealth*, 80.
149. See Miranda, *Communism*, 50f.
150. See Davids, *James*, 42.
151. Wolterstorff, *Justice*, 81. Wolterstorff adds: "No doubt this right, like others, can be forfeited; perhaps it is forfeited if a person refuses to work when decent work is available. And no doubt, as with other rights, there are social situations in which the right is abrogated—as, for example, when there are no arrangements that other parties can make to ensure our sustenance" (81). See also Wolterstorff, "Christianity."
152. Another way to stress the same point is to indicate that love is a necessary prerequisite of equality. Since "much of what makes our world . . . unsafe arises from economic and racial inequality" (Bellah, *Habits*, 285), struggle for equality is indispensable to making our world humane. Yet as much as equality is not only an inherent attribute of individuals (as in liberalism), it can also not be only an attribute of social justice (as in socialism). *Grace* is required to establish equality of possession without violating the equality of treatment.
153. Hay, "Order," 88.
154. See above, 59. In the sociological literature of the past two decades, we have witnessed a rediscovery of the Marxian critique of work as only means (see Seeman, "Alienation Motif," 179).
155. See above, 10ff.
156. See Korsgaard, "Goodness," 188.
157. See above, 127f.
158. It is not enough to say that human beings were created by God in such a way that they cannot live without working like they were created in such a way that they cannot live without breathing. Work is a positive purpose for which human beings were created; breathing is not. Work should, therefore, be (partly) an end in itself, whereas it would be absurd to claim the same of breathing.
159. Ryan, "Humanistic Work," 18. On enjoyment of work cf. Coomaraswamy, "A Figure of Speech," 41.
160. Luther, *WA*, XLII, 78, 4–5; cf. Calvin, *Genesis*, 125.
161. It is not clear to me how Weber can consider work done out of a sense of calling (as he defines it) to be "labour . . . performed as it were an absolute end in itself" (Weber, *Ethic*, 62). One of the main points of his book *The Protestant Ethic and the Spirit of Capitalism* is that the devotion to labor in the calling was precisely not an end in itself but a means to glorify God (108f.) and a means to show that one is in the state of grace (109ff.). To view work truly as an end in itself means to ascribe goodness to the action of working, pure and simple—to working that is, of course, by definition instrumental!—and to presuppose enjoyment in such action.

162. See above, 105ff.
163. See above, 58f.
164. For obvious reasons this does not apply to the types of work in which the internal goal of work is one with the activity of working (such as the work of a performing artist).
165. See Volf, *Zukunft*, 87f.
166. See De Man, *Work*, 39.
167. Durkheim, *Labour*, 371. Cf. Friedmann, *Work*, 32.
168. See Sölle, *Arbeiten*, 132.

Bibliography

The bibliography contains only the titles I refer to in the footnotes.

Acton, H. B. *Kant's Moral Philosophy*. London: Macmillan, 1970.
Agrell, G. *Work, Toil and Sustenance: An Examination of the View of Work in the New Testament, Taking into Consideration Views Found in Old Testament, Intertestamental, and Early Rabbinic Writings*. Lund: Håkan Ohlsons, 1976.
"All Work and No Play." In *Newsweek*, January 24, 1983: 20–25.
Alperovitz, G. "Planning for a Sustained Community." In *Catholic Social Teaching and the United States Economy: Working Papers for a Bishop's Pastoral*, ed. J. W. Houck and O. F. Williams, 331–58. Washington D. C.: University of America Press, 1984.
Altner, G. "Technisch-wissenschaftliche Welt und Shöpfung." In *Christlicher Glaube in moderner Gesellschaft*, ed. F. Böckle et al., 20: 86–118. Freiburg: Herder, 1982.
Ambler, R. *Global Theology: Faith in the Present World Crisis*. Philadelphia: Trinity Press International, 1990.
"America Runs Out of Time." In *Time*, April 24, 1989: 52–54.
Anthony, P. D. *Ideology of Work*. London: Tavistock, 1977.
Aquinas, T. *Summa Contra Gentiles*. Notre Dame: University of Notre Dame, 1975.
———. *Summa Theologica*. Westminster: Christian Classics, 1948.
———. *Questiones quodlibertales*. Ed. P. Fr. R. Spazzi, O. P. Roma: Marietti, 1956.

Arendt, H. *Vita activa oder Vom tätigen Leben.* München: R. Piper, 1981.
Aristotle. *Nicomachean Ethics.* In *The Complete Works of Aristotle,* ed. J. Barnes. Princeton: Princeton University Press, 1984.
———. *Politics.* Cambridge: Harvard University Press, 1932.
Arneson, R. J. "Meaningful Work and Market Socialism." In *Ethics* 97 (1987): 517-49.
Atra-Hasis: The Babylonian Story of the Flood. Ed. W. G. Lambert and A. R. Millard. Oxford: Clarendon Press, 1969.
Atteslander, P. "Von Arbeits- zur Tätigkeitsgesellschaft." In *Leben wir, um zu arbeiten? Die Arbeitswelt im Umbruch.* Ed. F. Niess, 125-33. Köln: Bund-Verlag, 1984.
Auer, A. *Umweltethik: Ein theologischer Beitrag zur ökologischen Diskussion.* Düsseldorf: Patmos, 1984.
Augustine. *The Confessions of St. Augustin.* In *A Select Library of the Nicene and Post-Nicene Fathers of the Christian Church,* ed. P. Schaff, 2:27-207. Grand Rapids: Eerdmans, 1956.
———. *Concerning the City of God Against the Pagans.* Harmondsworth: Penguin Books, 1972.
Balz, H. *Heilsvertrauen und Welterfahrung. Strukturen der paulinischen Eschatologie nach Römer 8,18-39.* BEvT 59. Göttingen: Vandenhoeck & Ruprecht, 1971.
Barnabas. "The Epistle of Barnabas." In *Apostolic Fathers,* 1:340-409. Cambridge: Harvard University Press; London: Heinemann, 1975.
Barrett, C. K. *A Commentary on the First Epistle to the Corinthians.* BNTC. New York: Harper & Row, 1968.
Barth, K. *Church Dogmatics.* Edinburgh: T. & T. Clark, 1936-1970.
Barth, M. *Ephesians.* Garden City: Doubleday, 1974.
Bayer, O. "Beruf." In *Evangelisches Soziallexikon,* 7th ed., ed. T. Schober et al., 140-42. Stuttgart: Kreuz, 1980.
———. "Tu dich auf! Verbum sanans und salvificans und das Problem der 'natürlichen' Theologie." In *Schöpfung als Anrede. Zu einer Hermeneutik der Schöpfung,* 62-79. Tübingen: Mohr, 1986.
Bebel, A. *Die Frau und der Sozialismus.* Stuttgart: Diez, 1913.
Bellah, R. N., et al. *Habits of the Heart: Individualism and Commitment in American Life.* New York: Harper & Row, 1985.
Benedict, "The Rule of Saint Benedict." In *Western Asceticism,* ed. O. Chadwick, 290-337. Philadelphia: Westminster Press, 1968.
Berger, K. "Charisma, ktl." In *EWNT,* ed. Horst Balz und Gerhard Schneider, 3:1102-1105. Stuttgart: Kohlhammer, 1983.
Berkhof, H. *Christ the Meaning of History.* Richmond: John Knox, 1966.
———. *The Doctrine of the Holy Spirit.* Richmond: John Knox Press, 1964.

Beversluis, E. H. "A Critique of Ronald Nash on Economic Justice and the State." In *Economic Justice and the State: A Debate Between Ronald H. Nash and Eric H. Beversluis*, ed. J. A. Bernbaum, 25–47. Grand Rapids: Baker Book House; Washington D.C.: Christian College Coalition, 1986.

Bielby, W. T., and J. N. Baron. "Men and Women at Work: Sex Segregation and Statistical Discrimination." In *AJS* 91 (1986): 759–99.

Bienert, W. *Die Arbeit nach der Lehre der Bibel. Eine Grundlegung evangelischer Sozialethik*. Stuttgart: Evangelisches Verlagswerk, 1954.

Biondić, I. *Specijalni odgoj na prekretnici. Prilog sociologiji odgoja i obrazovanja*. Zagreb: Institut za pedagogijska istraživanja Filozofskog fakulteta sveučilišta u Zagrebu, 1986.

Bloom, A. *The Closing of the American Mind: How Higher Education Has Failed Democracy and Impoverished the Souls of Today's Students*. New York: Simon and Schuster, 1987.

Bockmühl, K. "Protestant Ethics: The Spirit and the Word in Action." In *ERT* 12 (1988): 101–15.

Brakelmann, G. "Arbeit." In *Christlicher Glaube in moderner Gesellschaft*, ed. F. Böckle et al., 8:100–135. Freiburg: Herder, 1980.

Branda, L. *Work and Workers: A Sociological Analysis*. New York: Praeger, 1975.

Braverman, H. *Labor and Monopoly Capital: The Degradation of Work in the Twentieth Century*. New York: Monthly Review Press, 1974.

Brockhaus, H. *Charisma und Amt. Die paulinische Charismenlehre auf dem Hintergrund der frühchristlichen Gemeindefunktionen*. Wuppertal: Brockhaus, 1972.

Bruce, F. F. *The Epistle to the Galatians. A Commentary on the Greek Text*. NIGTC. Grand Rapids: Eerdmans, 1982.

———. *The Epistle of Paul to the Romans: An Introduction and Commentary*. Grand Rapids: Eerdmans, 1963.

Bultmann, R. "New Testament and Mythology." In *Kerygma and Myth: A Theological Debate*, ed. H. W. Bartsch, 1–44. New York: Harper & Row, 1961.

Calvin, J. *The First Epistle of Paul the Apostle to the Corinthians*. Grand Rapids: Eerdmans, 1960.

———. *Sermons on the Epistle to the Ephesians*. Edinburgh: Banner of Truth, 1973.

———. *Commentaries on the First Book of Moses Called Genesis*. Grand Rapids: Eerdmans, 1948.

———. *Institutes of the Christian Religion*. Ed. J. T. McNeill. Philadelphia: Westminster Press, 1977.

Carlyle, T. *Past and Present*. Boston: The Riverside Press, 1965.

Catherwood, Sir F. "Christian Faith and Economics." In *Transformation* 4, nos. 3/4 (1987): 1-6.
―――. "The New Technology: The Human Debate." In *The Year 2000*, ed. J. Stott, 126-45. Downers Grove: InterVarsity, 1983.
Chaw, P. "Technology and the Kingdom: An Approach to Evangelism in a Hungry World." In *Transformation* 4, no. 2 (1987): 16-20.
Chenu, M. D. *The Theology of Work: An Exploration*. Chicago: Regnery, 1966.
"20 to 200 Million Children Under 15 Are in World's Work Force." In *UNChron* 23, no. 5 (1986): 116.
Clement of Alexandria. *The Paedagogus*. In *The Ante-Nicene Fathers: Translation of the Writings of the Fathers Down to A.D. 325*, ed. A. Roberts and J. Donaldson, 2:207-98. Grand Rapids: Eerdmans, 1977.
―――. *The Salvation of the Rich Man. Who Is the Rich Man That Shall Be Saved?* In *The Ante-Nicene Fathers. Translation of The Writings of the Fathers Down to A.D. 325*, ed. A. Roberts and J. Donaldson, 2:589-604. Grand Rapids: Eerdmans, 1977.
Collste, G. "Toward a Normative Work Ethic." In *Will the Future Work? Values for Emerging Patterns of Work and Employment*, ed. H. Davis and D. Gosling, 94-100. Genève: WCC, 1984.
Congar, Y. *Der Heilige Geist*. Freiburg: Herder, 1982.
Cooley, M. *Architect or Bee? The Human/Technology Relationship*. Boston: South End Press, 1982.
Coomaraswamy, A. K. "A Figure of Speech or a Figure of Thought?" In *Selected Papers: Traditional Art and Symbolism*, ed. R. Lipsey, 13-42. Princeton: Princeton University Press, 1977.
Cranfield, C. E. B. *The Epistle to the Romans*. ICC. Edinburgh: T. & T. Clark, 1975.
Crosby, F., et al. "Cognitive Biases in the Perception of Discrimination: The Importance Format." In *Sex Roles* 14 (1986): 637-46.
Dabney, D. L. "Die Kenosis des Geistes: Kontinuität zwischen Schöpfung und Erlösung im Werk des Heiligen Geistes." Th. D. diss., University of Tübingen, 1989.
Dalferth, I. U., and E. Jüngel. "Person und Gottebenbildlichkeit." In *Christlicher Glaube in moderner Gesellschaft*, ed. F. Böckle et al., 24:57-99. Freiburg: Herder, 1981.
Davids, P. *Commentary on James: A Commentary on the Greek Text*. NIGTC. Grand Rapids: Eerdmans, 1982.
Descartes, R. *Discourse on Method; or, Rightly Conducting the Reason and Seeking Truth in the Sciences*. In *A Discourse on Method*. London: Dent & Sons; New York: Dutton, 1937.

———. *Meditations on the First Philosophy.* In op. cit.
Dickens, C. *Hard Times.* New York: Norton, 1966.
The Documents of Vatican II. Ed. W. M. Abbott, S.J. New York: Guild Press, 1966.
Van Drimmelen, R. "Homo Oikumenicus and Homo Economicus: Christian Reflection and Action on Economics in the Twentieth Century." In *Transformation* 4, nos. 3/4 (1987): 66–84.
Drucker, P. "Twilight of the First-Line Supervisor?" *The Wall Street Journal,* June 7, 1983: 100.
Durkheim, E. *The Division of Labour in Society.* Glencoe: Free Press, 1947.
Ebeling, G. *Luther: An Introduction to His Thought.* Philadelphia: Fortress Press, 1970.
Eckert, J. "Kaleō, ktl." In *EWNT,* ed. Horst Balz und Gerhard Schneider, 2:592–601. Stuttgart: Kohlhammer, 1981.
"Economic Justice for All: Catholic Social Teaching and the U.S. Economy." In *Origins* 16 (1986): 410–55.
Ellul, J. *The Technological Society.* New York: Vintage Books, 1964.
———. *The Technological System.* New York: Continuum, 1980.
Erickson, K. "On Work and Alienation." In *American Sociological Review* 51 (1986): 1–8.
Evangelism and Social Responsibility: An Evangelical Commitment. Exeter: Paternoster Press, 1982.
Evans, J. "Arbeitnehmer und Arbeitsplatz." In *Auf Gedeih und Verderb. Mikroelektronik und Gesellschaft. Bericht an den Club of Rome,* ed. G. Friedrich and A. Schaff, 169–200. Wien: Europaverlag, 1982.
Fee, G. D. *The First Epistle to the Corinthians.* Grand Rapids: Eerdmans, 1987.
Feuerbach, L. *Sämtliche Werke.* Ed. W. Bolin and F. Jodl. Stuttgart-Bad Cannstatt: Frommann, 1960.
Fichte, J. G. *Nachgelassene Werke.* Ed. I. H. Fichte. Bonn: A. Marcus, 1834–1835.
Field, D., and E. Stephenson. *Just the Job: Christians Talk about Work and Vocation.* Leicester: InterVarsity Press, 1978.
Fitzgerald, R. "Abraham Maslow's Hierarchy of Needs—An Exposition and Evaluation." In *Human Needs and Politics,* ed. R. Fitzgerald, 36–51. Oxford: Pergam, 1977.
Friedmann, G. *The Anatomy of Work: Labor, Leisure and the Implications of Automation.* Westport: Greenwood, 1962.
Friedman, M. *Capitalism and Freedom.* Chicago: University of Chicago Press, 1962.
Fullinwider, R. K. *The Reverse Discrimination Controversy. A Moral and Legal Analysis.* Totowa: Rowman and Littlefield, 1980.

Gandhi, M. K. *Bread Labour: The Gospel of Work.* Ed. R. Kelekar. Ahmedabad: Navajivan P. H., n. d.

Gatzen, H. "Beruf bei Martin Luther und in der industriellen Gesellschaft." Th. D. diss., University of Münster, 1964.

Gese, H. "Der Tod im Alten Testament." In *Zur Biblischen Theologie*, 31–54. München: Kaiser, 1977.

Gillett, R. W. *The Human Enterprise: A Christian Perspective on Work.* Kansas City: Leaven Press, 1985.

Ginzberg, E. "The Mechanization of Work." *Scientific American* 247 (September 1982): 67–75.

Gorbatschow, M. *Perestroika. Die zweite russische Revolution. Eine neue Politik für Europa und die Welt.* München: Knaur, 1987.

Gorz, A. *Wege ins Paradies. Thesen zur Krise, Automation und Zukunft der Arbeit.* Berlin: Rotbuch, 1983.

Griffits, B. *Morality in the Market Place: Christian Alternatives to Capitalism and Socialism.* Sevenoaks: Hodder & Stoughton, 1982.

———. *The Creation of Wealth. A Christian's Case for Capitalism.* Downers Grove: InterVarsity Press, 1984.

Günkel, H. *Genesis.* Göttingen: Vandenhoeck & Ruprecht, 1964.

Gundry, R. H. *Matthew: A Commentary on His Literary and Theological Art.* Grand Rapids: Eerdmans, 1985.

———. "The New Jerusalem. People as Place, Not Place for People (Revelation 21:1–22:5)." In *NovT* 29 (1987): 254–64.

Habermas, J. *Erkenntnis und Interesse.* Frankfurt am Main: Suhrkamp, 1979.

———. "Nachholende Revolution und linker Revisionsbedarf: Was heißt Sozialismus heute?" In *Die nachholende Revolution: Kleine Politische Schriften* VII, 179–204. Frankfurt am Main: Suhrkamp, 1990.

Hackman, J. R. "The Psychology of Self-Management in Organizations." In *Psychology and Work: Productivity, Change and Employment*, ed. M. S. Pallak and R. O. Perloff, 89–136. Washington: American Psychological Association, 1986.

Hall, R. H. *Dimensions of Work.* Beverly Hills: Sage Publications, 1986.

Harnack, A. von. *Die Mission und Ausbreitung des Christentums in den ersten drei Jahrhunderten*, 4th ed. Leipzig: J. C. Hinrichs'sche Buchhandlung, 1924.

Harper, M. *Let My People Grow: Ministry and Leadership in the Church.* London: Hodder & Stoughton, 1977.

Hartmann, K. "Was ist und was will Ethik? Ihre Herausforderung durch das naturwissenschaftlich und medizinisch Machbare." In *Concilium* (D) 25 (1989): 199–210.

Haughey, J. C. *Converting Nine to Five. A Spirituality of Daily Work.* New York: Crossroad, 1989.
Hawtrey, K. "Work and Leisure in Evangelical Focus." Paper presented as a part of a worldwide study process organized by Oxford Conference on Christian Faith and Economics. Australia, January, 1989.
Hay, D. A. *Economics Today: A Christian Critique.* Leicaster: Apollos, 1989.
──── . "The International Socio-Economic-Political Order and Our Lifestyle." In *Lifestyle in the Eighties: An Evangelical Commitment to Simple Lifestyle,* ed. R. J. Sider, 84–128. Philadelphia: Westminster, 1982.
──── . "North and South: The Economic Debate." In *The Year 2000,* ed. J. Stott, 72–102. Downers Grove: InterVarsity, 1983.
Hebblethwaite, B. *The Christian Hope.* Grand Rapids: Eerdmans, 1985.
Hegel, G. W. F. *Frühe politische Systeme,* ed. G. Göhler. Frankfurt am Main: Ullstein, 1974.
──── . *Werke,* ed. E. Moldenhauer and K. M. Michel. Frankfurt am Main: Suhrkamp, 1969–1970.
Heilbroner, R. L. *The Act of Work.* Washington: Library of Congress, 1985.
Hengel, M. "Die Arbeit im frühen Christentum." In *Theologische Beiträge* 17 (1986): 174–212.
Heron, A. I. C. *The Holy Spirit: The Holy Spirit in the Bible, the History of Christian Thought, and Recent Theology.* Philadelphia: Westminster Press, 1983.
Heschel, A. J. *The Sabbath: Its Meaning for Modern Man.* New York: Farrar, Straus and Giroux, 1980.
Hesiod, *Works and Days.* In *Hesiod: The Homeric Hymns and Homerica.* London: Heinemann; Cambridge: Harvard University Press, 1936.
Hill, S. *Competition and Control at Work: The New Industrial Sociology.* London: Heinemann; Cambridge: MIT Press, 1981.
Hoekema, A. A. *The Bible and the Future.* Grand Rapids: Eerdmans; Exeter: Paternoster, 1979.
Holmes, A. F. *Ethics: Approaching Moral Decisions.* Downers Grove: InterVarsity Press, 1984.
Honecker, M. "Die Krise der Arbeitsgesellschaft und das christliche Ethos." In *ZThK* 80 (1983): 204–22.
Ignatieff, M. *The Needs of Strangers: An Essay on Privacy, Solidarity, and the Politics of Being Human.* New York: Penguin Books, 1986.
Illich, I. *Selbstbegrenzung. Eine politische Kritik der Technik.* Reinbek bei Hamburg: Rowohlt, 1975.
Jacob, B. *Das erste Buch der Tora. Genesis.* Berlin: Schocken, 1934.

Jahoda, M. *Wieviel Arbeit braucht der Mensch? Arbeit und Arbeitslosigkeit im 20. Jahrhundert.* Weinheim: Beltz, 1983.

Janzen, K.-H. "Auswirkungen der neueren Technologien." In *Leben wir, um zu arbeiten? Die Arbeitswelt im Umbruch,* ed. F. Niess, 26–33. Köln: Bund-Verlag, 1984.

Jensen, O. "Schöpfungstheologischer Materialismus." In *NZSTh* 19 (1977): 247–60.

Joest, W. *Dogmatik I. Die Wirklichkeit Gottes.* Göttingen: Vandenhoeck & Ruprecht, 1984.

Johnson, P. G. *Grace: God's Work Ethic. Making Connections Between the Gospel and Weekday Work.* Valley Forge: Judson Press, 1985.

Johnston, R. K. *The Christian at Play.* Grand Rapids: Eerdmans, 1983.

Jonas, H. *Das Prinzip Verantwortung. Versuch einer Ethik für die technologische Zivilisation.* 3rd ed. Frankfurt am Main: Suhrkamp, 1982.

Jüngel, E. "Eschatologie. Thesen." Tübingen. Lectures handout.

———. *Zur Freiheit eines Christenmenschen. Eine Erinnerung an Luthers Schrift.* München: Kaiser, 1981.

Käsemann, E. "Amt und Gemeinde im Neuen Testament." In his *Exegetische Versuche und Besinnungen,* 1:109–34. Göttingen: Vandenhoeck & Ruprecht, 1970.

———. "Gottesdienst im Alltag der Welt." In op. cit., 2:198–204.

Kaiser, E. G. "Theology of Work." In *New Catholic Encyclopedia,* 14:1015–17.

Kant, I. *The Doctrine of Virtue. Part II of the Metaphysics of Morals.* New York: Harper & Row, 1963.

———. *Foundations of the Metaphysics of Morals.* New York: Bobbs-Merrill, 1969.

———. *Kritik der Urteilskraft.* Ed. K. Vorländer. Hamburg: Meiner, 1963.

Kasper, W. "Die Kirche als Sakrament der Geistes." In *Kirche—Ort des Geistes,* ed. W. Kasper and G. Stauter, 13–55. Freiburg: Herder, 1976.

Kern, W., and Y. Congar. "Geist und Heiliger Geist." In *Christlicher Glaube in moderner Gesellschaft,* ed. F. Böckle et al., 22:60–116. Freiburg: Herder, 1982.

King, A. "Einleitung. Eine neue industrielle Revolution oder bloß eine neue Technologie?" In *Auf Gedeih und Verderb. Mikroelektronik und Gesellschaft. Bericht an den Club of Rome,* ed. G. Friedrich and A. Schaff, 11–47. Wien: Europaverlag, 1982.

Kitagawa, J. "Reflections on the Work Ethic in the Religions of East Asia." In J. Pelikan, J. Kitagava, S. H. Nasr. *Comparative Work Ethics. Judeo-Christian, Islamic, and Eastern,* 27–47. Washington: Library of Congress, 1985.

Kluge, F. *Etymologisches Wörterbuch der deutschen Sprache.* Ed. W. Mitzka. Berlin: Walter de Gruyter, 1957.
Koch, K. "Gestaltet die Erde, doch heget das Leben! Einige Klarstellungen zum dominium terrae in Genesis 1." In *Wenn nicht jetzt, wann dann?" Aufsätze für Hans-Joachim Kraus zum 65. Geburtstag*, ed. H.-G. Geier et al., 23–36. Neukirchen-Vluyn: Neukirchener, 1983.
Kohn, M. L. "Occupational Structure and Alienation." In *Work and Personality: An Inquiry into the Impact of Social Stratification*, ed. M. L. Kohn and C. Schooler, 83–97. Norwood: Ablex, 1983.
Kohn, M. L., and C. Schooler. "The Reciprocal Effects of the Substantive Complexity of Work and Intellectual Flexibility: A Longitudinal Assessment." In *op. cit.*, 103–24.
Kolakowski, L. *Falls es keinen Gott gibt.* München: R. Pieper, 1982.
———. *Die Hauptströmungen des Marxismus. Entstehung, Entwicklung, Zerfall.* 2nd ed. München: R. Piper, 1981.
Kono, T. *Strategy and Structure of Japanese Enterprise.* Armouk: M. E. Sharpe, 1984.
Korsgaard, C. M. "Two Distinctions in Goodness." In *Philosophical Review* 92 (1983): 165–95.
———. "Kant's Formula of Humanity." In *Kant-Studien* 77 (1986): 183–202.
Kranzberg, M., and J. Gies. *By the Sweat of Thy Brow: Work in the Western World.* New York: Putnam's Sons, 1975.
Kraus, "Coming to Grips with the New Leisure." In *Leisure: No Enemy But Ignorance*, 11–15. Reston: The Alliance, 1983.
Kraus, H.-J. *Theologische Religionskritik.* Neukirchen-Vluyn: Neukirchner, 1982.
Krusche, W. *Das Wirken des Heiligen Geistes nach Calvin.* Göttingen: Vandenhoeck & Ruprecht, 1957.
Küng, H. *The Church.* New York: Doubleday, 1976.
———. *Projekt Weltethos.* München: Piper, 1990.
Kuzmič, P. "History and Eschatology: Evangelical Views." In *In Word and Deed. Evangelism and Social Responsibility*, ed. B. J. Nicholls, 135–64. Exeter: Paternoster Press, 1985.
Laborem Exercens. Encyclical Letter of the Supreme Pontiff John Paul II on Human Work. London: Catholic Truth Society, 1981.
Labour, Employment and Unemployment: An Ecumenical Reappraisal. Ed. R. H. Green. Geneva: WCC, 1987.
Ladd, G. E. *A Theology of the New Testament.* Grand Rapids: Eerdmans, 1974.
Lake, P. *Moderate Puritans and the Elizabethan Church.* Cambridge: Cambridge University Press, 1982.

Lamb, R. "Adam Smith's Concept of Alienation." In *Oxford Economic Papers* 25 (1973): 275-85.
Lampe, G. *God as Spirit*. London: SCM, 1983.
Landmann, M. *Fundamental-Anthropologie*. Bonn: H. Grundmann, 1979.
Langan, T. "The Changing Nature of Work in the World System." In *Communio* 11 (1984): 120-35.
Lasalett, P. "Introduction." In J. Locke, *Two Treatises of Government*, 3-152. Cambridge: Cambridge University Press, 1966.
Leiss, W. *The Domination of Nature*. New York: Georges Braziller, 1972.
Liedke, G. *Im Bauch des Fisches. Ökologische Theologie*. Stuttgart: Kreuz, 1979.
Lindgren, J. R. *The Social Philosophy of Adam Smith*. The Hague: Martinus Nijhoff, 1973.
Lochman, J. M. *Marx begegnen. Was Christen und Marxisten eint und trennt*. Gütersloh: Gütersloher Verlagshaus Mohn, 1975.
―――. "Werk und Werkgerechtigkeit. Arbeit in christlicher und marxistischer Sicht." In *ZEE* 22 (1978): 105-17.
Locke, J. *Two Treatises of Government*. Cambridge: Cambridge University Press, 1966.
Luther, M. *D. Martin Luther's Werke. Kritische Gesammtausgabe*. Weimar: H. Böhlau, 1883-.
Lyon, D. *The Silicon Society*. Grand Rapids: Eerdmans, 1986.
De Man, H. *Joy in Work*. London: George Allen & Unwin, 1929.
MacIntyre, A. C. *After Virtue: A Study in Moral Theory*. 2nd ed. Notre Dame: University of Notre Dame Press, 1984.
MacPherson, C. B. *The Political Theory of Possessive Individualism: Hobbes to Locke*. Oxford: Clarendon Press, 1962.
Marx, K. *Grundrisse der Kritik der politischen Ökonomie*. Berlin: Diez, 1974.
Marx, K. *Resultate des unmittelbaren Produktionsprozesses. Das Kapital. I. Buch. Der Produktionsprozeß des Kapitals. VI. Kapitel*. Frankfurt am Main: Suhrkamp, 1969.
Marx, K., and F. Engels. *Marx-Engels Werke*. Berlin: Diez, 1972.
―――. *Marx-Engels Werke. Ergänzungsband*. Berlin: Diez, 1968.
Maslow, A. H. *Motivation and Personality*. 2nd ed. New York: Harper & Row, 1970.
Mater et Magistra. Johannes XXIII. In *Die sozialen Enzykliken*, ed. J. Binkowski, 91-150. Villingen: Ring, 1963.
McNulty, P. J. "Adam Smith's Concept of Labor." In *Journal of the History of Ideas* 34 (1973): 345-66.
McPherson, J. "Towards an Ecological Theology." In *ET*, 97 (1985-86): 236-42.

Meek, R. L. *Smith, Marx and After: Ten Essays in the Development of Economic Thought*. London: Chapman & Hall, 1977.

Meeks, M. D. *God the Economist: The Doctrine of God and Political Economy*. Philadelphia: Fortress Press, 1989.

Mészáros, I. *Marx's Theory of Alienation*. 4th ed. London: Merlin Press, 1975.

Mieth, D. *Arbeit und Menschenwürde*. Freiburg: Herder, 1985.

——— . *Einheit von vita activa und vita contemplativa in den deutschen Predigten und Traktaten Meister Eckharts und bei Johannes Tauler*. Regensburg: Verlag Friedrich Puset, 1969.

Mill, J. S. *Utilitarianism*. In *Collected Works*, ed. J. M. Robson, 10:205–59. Toronto: University of Toronto Press, 1969.

Miranda, J. P. *Communism in the Bible*. New York: Orbis Books, 1982.

Von Mises, L. "Liberalismus II. Wirtschaftlicher Liberalismus." In *HdSW*, ed. E. von Beckerath et al., 6:596–603. Stuttgart: Gustav Fischer, 1959.

Moltmann, J. "Christsein, Menschsein und das Reich Gottes. Ein Gespräch mit Karl Rahner." In *Stimmen der Zeit* 203 (1985): 619–31.

——— . *God in Creation: A New Theology of Creation and the Spirit of God*. San Francisco: Harper & Row, 1985.

——— . "The Right to Work." In *On Human Dignity: Political Theology and Ethics*. Philadelphia: Fortress Press, 1984.

——— . *Theology of Hope: On the Ground and the Implications of a Christian Eschatology*. New York: Harper & Row, 1967.

——— . *Trinität und Reich Gottes*. München: Kaiser, 1980.

——— . *Der Weg Jesu Christi. Christologie in messianischen Dimensionen*. München: Kaiser, 1989.

Mounce, R. H. *The Book of Revelation*. Grand Rapids: Eerdmans, 1977.

Mühlen, H. "Charisma und Gesellschaft." In *Geistesgaben heute*, ed. H. Mühlen, 160–74. Mainz: Matthias-Grünewald, 1982.

Munroe, R. H., R. L. Munroe, H. S. Shimmin. "Children's Work in Four Cultures: Determinants and Consequences." In *American Anthropologist* 86 (1984): 369–79.

Naisbitt, J. *Megatrends. Ten New Directions Transforming Our Lives*. London: Macdonald, 1984.

Naisbitt, J., and P. Aburdene. *Reinventing the Corporation: Transforming Your Job and Your Company for the New Information Society*. New York: Warner Books, 1985.

Nash, R. H. "A Reply to Eric Beversluis." In *Economic Justice and the State: A Debate Between Ronald H. Nash and Eric H. Beversluis*, ed. J. A. Bernbaum, 49–65. Grand Rapids: Baker Book House; Washington D.C.: Christian College Coalition, 1986.

Nash, R. *Social Justice and the Christian Church.* Milford: Mott Media, 1983.
Neff, W. S. *Work and Human Behavior.* 2nd ed. Chicago: Aldim, 1977.
Nehamas, A. "How One Becomes What One Is." In *The Philosophical Review* 92 (1983): 385–417.
Neulinger, J. *The Psychology of Leisure: Research Approaches to the Study of Leisure.* Springfield, Ill.: Charles C. Thomas, 1974.
Von Nell-Breuning, O. *Arbeitet der Mensch zu viel?* Freiburg: Herder, 1985.
———. "Kommentar." In Johannes Paulus II, *Über die menschliche Arbeit*, 103–27. Freiburg: Herder, 1981.
Niebuhr, R. *The Children of Light and the Children of Darkness: A Vindication of Democracy and Critique of its Traditional Defense.* New York: Charles Scribner's Sons, 1945.
Nielsen, K. "True Needs, Rationality and Emancipation." In *Human Needs and Politics*, ed. R. Fitzgerald, 142–156. Oxford: Pergamon Press, 1977.
Nietzsche, F. *Zur Genealogie der Moral.* In *Werke. Kritische Gesamtausgabe*, ed. G. Colli and M. Montinari, Vol. 6/2. Berlin: Walter de Gruyter, 1968.
Norman, C. *The God That Limps: Science and Technology in the Eighties.* New York: Norton, 1981.
North, R. *Sociology of the Biblical Jubilee.* Rome: Pontifical Biblical Institute, 1954.
Novak, M. *Toward a Theology of the Corporation.* Washington: American Enterprise Institute for Public Policy Research, 1981.
O'Donovan, O. *Resurrection and Moral Order: An Outline for Evangelical Ethics.* Leicester: InterVarsity; Grand Rapids: Eerdmans, 1986.
Origen. *Origen Against Celsus.* In *The Ante-Nicene Fathers. Translation of the Writings of the Fathers Down to A.D. 325*, ed. A. Roberts and J. Donaldson, 4:395–669. Grand Rapids: Eerdmans, 1979.
Ouellette, R. P., et al. *Automation Impacts on Industry.* Ann Arbor: Ann Arbor Science, 1983.
"The Oxford Conference on Christian Faith and Economics." In *Transformation* 4, no. 2 (1987): 22–24.
The Oxford Conference (Official Report). Ed. J. H. Oldham. Chicago: Wallett, Clark, 1937.
"The Oxford Declaration on Christian Faith and Economics." In *Transformation* 7, no. 3 (1990): 1–9.
Pannenberg, W. *Anthropologie in theologischer Perspektive.* Göttingen: Vandenhoeck & Ruprecht, 1983.
Parker, S. *The Future of Work and Leisure.* New York/Washington: Praeger, 1971.

Passmore, J. *Man's Responsibility for Nature.* London: Duckworth, 1974.
Paulk, E. P. *Your Pentecostal Neighbor.* Cleveland: Pathway Press, 1958.
"People Development Strategy." Paper of ServiceMaster Industries, 1976.
Pieper, J. *Leisure: The Basis of Culture.* New York: New American Library, 1963.
Pinnock, C. H. "Introduction." In *The Holy Spirit. Renewing and Empowering Presence*, ed. G. Vandervelde, 7–9. Winfield: Wood Lake Books, 1989.
Plato. *Laws.* London: Heinemann; Cambridge: Harvard University Press, 1926.
―――. *Republic.* New York: Modern Library, 1982.
―――. *Lysis, Symposium, Georgias.* Cambridge: Harvard University Press; London: Heinemann, 1939.
Pohl, R. E. *Divisions of Labour.* Oxford: Basil Blackwell, 1984.
Pomian, K. "Die Krise der Zukunft." In *Über die Krise. Castelgondolfo-Gespräche 1985*, ed. K. Michalski, 105–26. Stuttgart: Klett-Cotta, 1986.
Potter, N. "Kant on Ends That Are at the Same Time Duties." In *Pacific Philosophical Quarterly* 66 (1985): 78–92.
Preston, "Vocation." In *A Dictionary of Christian Ethics.* Ed. J. Macquarrie. London: SCM, 198–99.
Preuß, H. D. "Arbeit I." In *TRE*, ed. G. Krause and G. Müller, 3:613–18. Berlin: Walter de Gruyter, 1978.
Quadragessimo Anno. Pius XI. In *Die sozialen Enzykliken*, ed. J. Binkowski, 37–89. Villingen: Ring, 1963.
Von Rad, G. *Genesis. A Commentary.* Philadelphia: Westminster, 1972.
―――. *Theologie des Alten Testaments I. Die Theologie der geschichtlichen Überlieferungen Israels.* München: Kaiser, 1969.
Raines, J. C., and D. C. Day-Lower. *Modern Work and Human Meaning.* Philadelphia: Westminster Press, 1986.
Rasmussen, W. D. "The Mechanization of Agriculture." In *Scientific American* 247 (September 1982): 77–89.
Ratschow, C. H. "Eschatologie VIII." In *TRE*, ed. G. Krause and G. Müller, 10:334–63. Berlin: Walter de Gruyter, 1982.
Reisman, D. A. *Adam Smith's Sociological Economics.* London: Croom Helm; New York: Barnes & Noble, 1976.
Rerum Novarum. Leo XIII. In *Die sozialen Enzykliken*, ed. J. Binkowski, 1–35. Villingen: Ring, 1963.
Rich, A. "Arbeit als Beruf. Das christliche Verständnis der Arbeit." In *Arbeit und Humanität*, ed. A. Rich and E. Urlich, 7–19. Königstein Ts.: Athenäum, 1978.

———. *Wirtschaftsethik. Grundlagen in theologischer Perspektive.* Gütersloch: Gütersloher Verlaghaus Mohn, 1984.
Richardson, A. *The Biblical Doctrine of Work.* London: SCM Press, 1952.
Ricoeur, P. "Ist 'die Krise' ein spezifisch modernes Phänomen?" In *Über die Krise. Castelgondolfo-Gespräche 1985*, ed. K. Michalski, 38–63. Stuttgart: Klett-Cotta, 1986.
Rights of Future Generations—Rights of Nature: Proposal for Enlarging the Universal Declaration of Human Rights. Ed. L. Vischer, Studies from the World Alliance of Reformed Churches, no. 19, 1990.
Rinklin, A. "Mendevilles Bienenfabel: Private Laster als Quelle des Gemeinwohls." In *ZEE* 29 (1985): 216–28.
Rondet, H. *Die Theologie der Arbeit.* Würzburg: Echter, 1954.
Roos, L. "On a Theology and Ethics of Work." In *Communio* 11 (1984): 100–19.
Ruskin, J. *The Crown of Wild Olive: Four Lectures on Work, Traffic, War, and the Future of England.* New York: Thomas Y. Crowell, n.d.
Ryan, J. J. "Humanistic Work: Its Philosophical and Cultural Implications." In *A Matter of Dignity: Inquiry into the Humanization of Work*, ed. W. J. Heisler and J. W. Hauck, 11–22. Notre Dame: University of Notre Dame Press, 1977.
Sahlins, M. *Stone-Age Economics.* London: Tavistock, 1974.
Samuel, V., and C. Sugden. "Evangelism and Social Responsibility—A Biblical Study in Priorities." In *In Word and Deed: Evangelism and Social Responsibility*, ed. B. J. Nicholls, 198–214. Exeter: Paternoster Press, 1985.
Scheler, M. "Arbeit und Ethik." In his *Christentum und Gesellschaft*, 27–48. Leipzig: Der Neue Geist-Verlag, 1924.
Schmidt, W. H. *Der Schöpfungsbericht der Priesterschrift.* 3rd ed. Neukirchen-Vluyn: Neukirchener, 1973.
Schrage, W. "Heil und Heilung im Neuen Testament." In *EvTh* 46 (1986): 197–214.
Schulz, S. "Charismenlehre des Paulus. Bilanz der Probleme und Ergebnisse." In *Rechtfertigung. Festschrift für Ernst Käsemann zum 70. Geburtstag*, ed. J. Friedrich et al., 443–60. Göttingen: Vandenhoeck & Ruprecht; Tübingen: Mohr, Siebeck, 1975.
Schumacher, E. F. *Good Work.* New York: Harper & Row, 1979.
———. *Small Is Beautiful: A Study of Economics as If People Mattered.* London: Blond & Briggs, 1973.
Schumpeter, J. A. *History of Economic Analysis.* New York: Oxford University Press; London: Allen & Unwin, 1954.
Schweizer, E. *Heiliger Geist.* Stuttgart: Kreuz, 1978.
Scruton, R. *Kant.* Oxford: Oxford University Press, 1982.

Seeman, M. "Alienation Motif in Contemporary Theorizing: The Hidden Continuity of the Classic Themes." In *Social Psychology Quarterly*, 46 (1983): 171–84.

Shallis, M. *The Silicon Idol: The Microrevolution and Its Social Implications*. New York: Schocken Books, 1984.

Shelly, J. A. *Not Just a Job: Serving Christ in Your Work*. Downers Grove: InterVarsity, 1985.

Simon, Y. R. *Work, Society and Culture*. New York: Fordham University Press, 1971.

Smith, A. *An Early Draft of Part of the Wealth of Nations*. In W. R. Scott, *Adam Smith as Student and Professor*. Glasgow: Jackson, 1937.

——. *Lectures on Justice, Police, Revenues and Arms*. Ed. E. Cannan. Oxford: Clarendon Press, 1896.

——. *An Inquiry into the Nature and Causes of the Wealth of Nations*. New York: Random House, 1937.

——. *The Theory of Moral Sentiments*. New York: A. M. Kelley, 1966.

Smith, M. B. "Metapsychology, Politics, and Human Needs." In *Human Needs and Politics*, ed. R. Fitzgerald, 124–41. Oxford: Pergamon, 1977.

Sölle, D. *Lieben und Arbeiten. Eine Theologie der Schöpfung*. Stuttgart: Kreuz, 1985.

Solomon, K. *Die Lösung des sozialen Problems: die Bibel*. Breslau: Marcus, 1931.

Spenner, K. I. "Deciphering Prometheus: Temporal Change in the Skill Level of Work." In *American Sociological Review* 48 (1983): 824–37.

Spiegel, H. W. "Adam Smith's Heavenly City." In *Adam Smith and Modern Political Economy: Bicentennial Essays on the Wealth of Nations*, ed. G. P. O'Driscoll, Jr., 102–14. Ames: Iowa State University Press, 1979.

Steck, O. H. *Welt und Umwelt*. Stuttgart: Kohlhammer, 1978.

Stott, J. *Issues Facing Christians Today: A Major Appraisal of Contemporary Social and Moral Questions*. Basingstoke: Marshall Pickering, 1984.

Stout, J. *Ethics After Babel: The Language of Morals and Their Discontents*. Boston: Beacon Press, 1988.

Taylor, F. W. *The Principle of Scientific Management*. In his *Scientific Management*. New York: Harper & Row, 1947.

——. *Shop Management*. In op. cit.

Taylor, J. V. *The Go-Between God. The Holy Spirit and the Christian Mission*. London: SCM, 1972.

Tertullian, "On Idolatry." In *The Ante-Nicene Fathers: Translation of the*

Writings of the Fathers Down to A.D. *325*, ed. A. Roberts and J. Donaldson, 3:61-77. Grand Rapids: Eerdmans, 1976.

Thompson, E. P. *The Making of the English Working Class.* 2nd ed. Harmondsworth: Penguin, 1968.

Tödt, H. "Die Ambivalenz des technischen Fortschritts als Thema christlicher Ethik." In *ZEE* 25 (1981): 187-201.

Trilhaas, W. *Ethik.* 3rd ed. Berlin: Walter de Gruiter, 1970.

Underhill, E. *Worship.* New York: Harper, 1957.

Ure, A. *Philosophy of Manufactures.* London: Knight, 1835.

Veenhof, T. "Charismata—Supernatural or Natural?" In *The Holy Spirit: Renewing and Empowering Presence*, ed. G. Vandervelde, 73-91. Winfield: Wood Lake Books, 1989.

Volf, M. "Doing and Interpreting: An Examination of the Relationship Between Theory and Practice in Latin American Liberation Theology." In *Themelios* 8, no. 3 (1983): 11-19.

———. *I znam da sunce ne boji se tame. Teološke meditacije o Šantićevu vjerskom pjesništvu.* Zagreb: Izvori, 1986.

———. "Kirche als Gemeinschaft. Ekklesiologische Überlegungen aus freikirchlicher Perspektive." In *EvTh* 49 (1989): 52-76.

———. "Das Marxsche Verständnis der Arbeit. Eine theologische Wertung." Th. D. diss., University of Tübingen, 1985.

———. "Materiality of Salvation. An Investigation in the Soteriologies of Liberation and Pentecostal Theologies." In *Journal of Ecumenical Studies* 26 (1989): 447-67.

———. "On Human Work: An Evaluation of the Key Ideas of the Encyclical *Laborem exercens*." In *SJTh* 37 (1984): 67-79.

———. "On Loving with Hope: Eschatology and Social Responsibility." In *Transformation* 7, no. 3 (1990): 28-31.

———. *Zukunft der Arbeit—Arbeit der Zukunft. Der Arbeitsbegriff bei Karl Marx und seine theologische Wertung.* München: Kaiser; Mainz: Grünewald, 1988.

Wagner, F. "Berufung III. Dogmatisch." In *TRE*, ed. G. Krause and G. Müller, 5:688-713. Berlin: Walter de Gruyter, 1980.

Wallraff, G. *Ganz Unten.* Köln: Kiepenheuer & Witsch, 1985.

Weber, M. *The Protestant Ethic and the Spirit of Capitalism.* New York: Charles Scribner's Sons, 1958.

Welker, M. "'Unity of Religious History' and 'Universal Self-Consciousness': Leading Concepts or Mere Horizons on the Way *Towards a World Theology*?" In *HTR* 81 (1988): 431-44.

Welty, E. *Vom Sinn und Wert der menschlichen Arbeit.* Heidelberg: Kerle, 1949.

BIBLIOGRAPHY

West, E. G. *Adam Smith.* New Rochelle: Arlington House, 1969.
Westermann, C. *Genesis I–II: A Commentary.* Minneapolis: Augsburg, 1984.
Westermann, C. *Schöpfung.* Stuttgart: Kreuz, 1979.
Wilckens, U. *Der Brief an die Römer.* Zürich: Benzinger; Neukirchen-Vluyn: Neukirchener, 1980.
Williams, S. "The Partition of Love and Hope: Eschatology and Social Responsibility." In *In Transformation* 7, no. 3 (1990): 24–27.
Wingren, G. "Beruf II. Historische und ethische Aspekte." In *TRE*, ed. G. Krause and G. Müller, 5:657–71. Berlin: Walter de Gruyter, 1980.
Wingren, G. *Luther's Lehre vom Beruf.* München: Kaiser, 1952.
Wogeman, P. C. *The Great Economic Debate: An Ethical Analysis.* Philadelphia: Westminster Press, 1977.
Wogaman, P. C. *Economics and Ethics: A Christian Inquiry.* Philadelphia: Fortress Press, 1986.
Wolf, E. *Sozialethik. Theologische Grundfragen.* Göttingen: Vandenhoeck & Ruprecht, 1975.
Wolff, H. W. *Anthropologie des Alten Testaments.* Berlin: Evangelische Verlagsanstalt, 1980.
Wolterstorff, N. "The Bible and Economics: The Hermeneutical Issues." In *Transformation* 4, nos. 3/4 (1987): 11–19.
———. "Christianity and Social Justice." In *Christian Scholar's Review* 16 (1987): 211–28.
———. "Evangelicalism and the Arts." In *Christian Scholar's Review* 17 (1988): 449–473.
———. *Until Justice and Peace Embrace.* Grand Rapids: Eerdmans, 1983.
Wunsch, G. *Evangelische Wirtschaftsethik.* Tübingen: Mohr, 1927.
Xenophon. *Cyropaedia.* London: Heinemann; New York: Macmillan, 1925.
———. *Oeconomicus.* In *Memorabilia and Oeconomicus*, 361–532. London: Heinemann; Cambridge: Harvard University Press, 1952.
Yoder, J. H. *The Politics of Jesus.* Grand Rapids: Eerdmans, 1972.
Zimmerli, W. "Der Mensch im Rahmen der Natur nach den Aussagen des ersten biblischen Schöpfungsberichtes." In *ZThK* 76 (1979): 139–58.
———. "Mensch und Arbeit im Alten Testament." In *Recht auf Arbeit—Sinn der Arbeit*, ed. J. Moltmann, 40–58. München: Kaiser, 1979.

Index of Scriptural References

OLD TESTAMENT

Genesis
1	74, 77, 147
1:26	127
1:26–28	142, 146, 147
1:28	147
1:29	147
2	74, 77, 119, 128, 217n.16
2:5	99, 127
2:15	71, 119, 127–28, 145, 168
3	119, 217n.16
3:17ff.	119, 127–28, 167–68
3:23	127
4:1ff.	182
4:17ff.	93, 181
4:21f.	182
4:23f.	183
4:25ff.	181
9:9f.	145
9:10ff.	96
11:1–11	181–82

Exodus
1:13–14	164
5:9	222n.25
6:9	166
20:9	140
22:26–27	211n.33
31:2f.	130
35:2–3	113

Leviticus
25:8ff.	173
25:39ff.	164

Numbers
32:20–32	147

Deuteronomy
26:6–8	164

Joshua
18:1f.	147

INDEX OF SCRIPTURAL REFERENCES

Judges
3:10	114

1 Samuel
16:13	114
23:2	114

1 Chronicles
22:18f.	147
28:11–12	113

Job
34:14f.	144

Psalms
65:11–13	99
104	143, 147
104:23	128
104:29–30	144
112:9	194
127:1	99

Proverbs
16:10	114

Ecclesiastes
4:4	121
6:19	159

Isaiah
11:6–8	148, 154
11:6–10	95
14:14	182
28:24–29	114
37:26ff.	99
55:2	153
58:6	165
65:17–25	95
65:21	165
65:23	165
65:25	148, 154

Jeremiah
22:13	164

Ezekiel
16:49	194

Amos
9:13	165

NEW TESTAMENT

Matthew
5:5	94
6:1	194
6:6	137
6:10	94, 100
6:33	94, 133, 154, 201
9:37f.	93
11:19	151
12:28	102
19:28	213n.24
25:34	99
25:34ff.	93
25:40	213n.12
26:52	72

Mark
1:13	148
8:36	92, 133, 139
10:45	187

Luke
4:18	165
10:38ff.	72
10:42	152

INDEX OF SCRIPTURAL REFERENCES

12:15	72	5:14	125
12:34	210n.15	9:9	194
		13:13	193

John
1:12	102
17:21	137

Acts
2:17ff.	112
3:19–21	213n.24
20:35	139, 189, 195

Romans
8:18	16, 121
8:19–22	144, 154
8:21	95–96, 145–46
8:23	102, 115
11:29	116
12	214n.60
12:1f.	137
12:7	138
12:8	111

1 Corinthians
1:9	137
1:26	110
3:12–15	120
3:15	121
7:20	109–10
12:11	115–16
12:13	137, 153
12:28	138
12:31	116
12:44ff.	190
13:12	154
14:1	116
14:12	116–17, 131
14:12ff.	190
14:12–26	190

2 Corinthians
1:22	102, 115

Galatians
2:20	114, 141
5:13	186
5:22f.	131, 141

Ephesians
2:15ff.	191
4:11	111
4:13–16	191
4:24	191
4:28	72, 93, 139, 149, 187, 189
5:25–28	192
5:30	214n.53
6:6	125
6:8	97

Colossians
3:14	154
3:23	125

1 Thessalonians
4:11	72
4:12	149

2 Thessalonians
3:6ff.	93
3:10–12	192
3:10	26, 72, 139
3:12	189

1 Timothy
4:4	96
6:17	151

2 Timothy
1:6	131

James
1:28   165

1 Peter
1:4   99
1:15   110
2:9   110
2:18ff. 166
2:20   167
4:10f.   190

2 Peter
3:10   89

3:11   95
3:12   100

Revelation
2:9   94
14:13   97
21:2   99
21:22   137
21:24–26 213n.24
21:24   118
21:26   118
21:27   120
22:17   100

Index of Authors

Arendt, H., 226n.124
Aristotle, 48, 126, 136, 138
Augustine, 7, 139

Bacon, F., 148
Baron, J. N., 39
Barth, K., 73, 90, 191
Bellah, R., 3–4
Bentham, J., 190
Bielby, W. T., 39
Bruce, F. F., 95
Bultmann, R., 214n.49

Calvin, J., 105, 107, 117, 144, 198, 215n.79, 217n.18
Carlyle, T., 126–27
Chenu, M.-D., 71
Clement of Alexandria, 71, 210n.15

De Man, H., 200
Dostoyevsky, F., 80
Durkheim, E., 200–1

Fichte, J. G., 49, 209n.98
Friedman, M., 47

Gorbachev, M., 16
Gundry, R. H., 94
Gutenberg, J., 97

Hailey, A., 39
Hawtrey, K., 219n.73
Hay, D. A., 205n.49
Hayek, F. A., 47
Hegel, G. W. F., 49, 64, 132, 209n.98, 221n.115
Heilbroner, R. L., 11
Hesiod, 126

Ignatieff, M., 151

John Paul II, 5, 142

Kant, I., 170–73
Käsemann, E., 111
Kohn, M. L., 158, 218n.27
Kuhn, T., viii

251

Leo XIII, 5
Locke, J., 48, 51–52, 190
Luther, M., 73, 100, 103–10, 117, 145, 148, 156, 215n.66

MacIntyre, A., 187
Marx, K., 26, 46–50, 55–65, 132, 141, 145, 149–50, 157, 162–63, 170–72, 174, 181, 193, 196, 200, 204n.43, 207n.2, 209n.95
Moltmann, J., 77, 79, 216n.103

Nietzsche, F., 80, 132
Noyce, R., 34

Plato, 31, 51, 126, 143, 151, 214n.61

Rondet, H., 97

Santic, A., 27
Schooler, C., 218n.27
Schumpeter, J. A., 51
Smith, A., x, 26, 32, 46–55, 59–61, 130, 145, 149–50, 162, 174–75, 187–88, 196, 206n.21, 207n.2, 208n.49, 209n.98
Spenser, H., 47

Taylor, F. W., 174–75
Thomas Aquinas, 70, 90, 95, 216n.108

Weber, M., 106, 112, 227n.161
Welker, M., 217n.21
Wolterstorff, N., 139, 227n.151

Xenophon, 31